THE HEALTHY FAMILY
COOKBOOK

FAVORITE RECIPES® OF
HOME ECONOMICS TEACHERS

Great American Opportunities, Inc./Favorite Recipes Press

President: Thomas F. McDow III
Director of Marketing: Roger Conner
Marketing Services Manager: Karen Bird

Editorial Manager: Mary Jane Blount
Editors: Georgia Brazil, Mary Cummings, Jane Hinshaw,
Linda Jones, Mary Wilson
Typography: William Maul, Sharon Whitehurst

Home Economics Advisory Board

Favorite Recipes Press wants to recognize the following persons who graciously serve on our Home Economics Advisory Board:

Marilyn Butler
Home Economics Teacher
Midwest City, Oklahoma

Kathryn Jensen
Home Economics Teacher
Brigham City, Utah

Sandra Crouch
Home Economics Teacher
Sparta, Tennessee

Alma Payne
Home Economics Teacher
Hurst, Texas

Regina Haynes
Home Economics Teacher
Waynesville, North Carolina

Sherry Ziegler
Home Economics Teacher
Chillicothe, Ohio

Sue Smith
State Specialist, Home
Economics Education
Alabama Dept. of Education

Photography: Alaska Seafood Marketing Institute; Hershey Foods Corporation; Florida Department of Citrus; Tuna Research Foundation; Best Foods, Division of Corn Products Company; Louisiana Yam Commission; and Oregon-Washington-California Pear Bureau.

Library of Congress Catalog Number: 87-8451

ISBN: 0-87197-220-4

Manufactured in the United States of America

First Printing 1987
Second Printing 1988
Third Printing 1988

Your HEALTHY FAMILY Cookbook is designed to help build an awareness of the nutritional value in your daily meals. It is not a substitute for a carefully planned weight loss program or a medically supervised special diet. It is a guide to cooking and serving foods that are the most nutritionally beneficial to your family.

The editors have attempted to present these tried-and-true family recipes in a form that allows approximate nutritional values to be computed. Persons with dietary or health problems or whose diets require close monitoring should not rely solely on the nutritional information provided. They should consult their physicians or a registered dietitian for specific information.

Abbreviations for Nutritional Analysis

Cal—Calories   Sod—Sodium
Prot—Protein   Potas—Potassium
T Fat—Total Fat   gr—gram
Chol—Cholesterol   mg—milligram
Carbo—Carbohydrates

Nutritional information for recipes is computed from values furnished by the United States Department of Agriculture Handbook. Many specialty items and new products now available on the market are not included in this handbook. However, producers of new products frequently publish nutritional information on each product's packaging and that information may be added, as applicable, for a more complete analysis.

Unless otherwise specified, the nutritional analysis is based on the following guidelines:

- All measurements are level.
- Cottage cheese is cream-style with 4.2% creaming mixture. Dry curd has no creaming mixture.
- Eggs are large.
- Evaporated milk is canned milk produced by the removal of approximately 60% of the water from whole milk.
- Flour is unsifted all-purpose flour.
- Garnishes, serving suggestions and other essentially optional additions and variations are not included in the analysis.
- Herbs, spices, seasoning mixes, flavoring extracts and artificial sweeteners are not included in the analysis.
- Margarine and butter are regular, not whipped or presoftened.
- Milk is whole milk, 3.5% butterfat. Low-fat milk is 1% butterfat.
- Oil is any cooking oil, usually a blend of oils from several sources.
- Salt to taste as noted in the method has not been included in the nutritional analysis.
- Shortening is hydrogenated vegetable shortening for all-purpose use.

CONTENTS

The Healthy Family—It's Your Choice

We have all heard the expression, "You are what you eat." And with today's emphasis on fitness and nutrition, choosing foods for over 2000 meals and snacks that we eat each year is even more important than ever. Making healthful choices is one of the best ways we have of being good to ourselves and to our families.

However, each member of the family has different nutritional needs based on age, sex, activity level, and special health problems. Women, for example, need extra calcium to counteract loss of bone calcium; athletes may require higher levels of carbohydrates to ensure peak performance; joggers may require additional iron. According to the Senate Select Committee on Nutrition and Human Needs, all of us should **reduce** the levels of saturated and total fat, cholesterol, sodium, refined sugar, alcohol, and caffeine while **increasing** the amount of dietary fiber in the foods we eat.

The nutritional analysis provided with each recipe in *The Healthy Family Cookbook* will be very helpful in choosing foods for your family. Not only can you tell at a glance which recipes would be appropriate for diets high or low in calories, restricted in sodium or fat, or high in carbohydrates or potassium, but you can also add the nutritional elements for each menu together to ensure that all the family's needs are met daily.

Begin with Breakfast

One of the most important yet frequently overlooked ways to keep your family healthy and full of energy is to start the day with a nutritious breakfast. Sixty million Americans eat only two meals a day—lunch and dinner. For many others, breakfast consists of little more than coffee and a sweet roll. Yet, breakfast is the most important meal of the day for maintaining high energy levels, peak performance, and optimum use of calories. The good breakfast provides ¼ to ⅓ of the day's total calorie requirements and ¼ of the daily protein allowance. Plan breakfast menus using suggestions from the "Healthy Breakfasts" section of *The Healthy Family Cookbook*. Then supplement with fresh fruits, low-fat milk products, or whole-grain breads and cereals. You will eat fewer snacks, increase efficiency, decrease fatigue, and maintain weight more easily. For better breakfasts, follow these suggestions:

- Plan breakfast menus ahead.
- Prepare breakfast the night before if possible.
- Allow enough time in the morning to enjoy your food.
- Use a variety of recipes from day to day.
- Serve foods which are not usually breakfast foods such as sandwiches, pizza, or casseroles.
- Eat less dinner in order to be ready for breakfast.
- Switch breakfast and supper menus.

Make the Healthy Choice

For every meal and snack, make food choices which are good for you and your family.

Healthy Choice: Reduce refined sugar in your diet

- The average American consumes 39 teaspoons of sugar per day or 130 pounds per year.
- Watch for hidden sources of sugar in soft drinks, cakes and cookies, candy, chewing gum, canned fruits and juices, ice cream, jams and jellies, and syrups.
- Choose low-sugar snacks such as bagels, crackers, yogurt, and fresh fruit.

Healthy Choice: Reduce the total amount of fat and cholesterol in your diet

- Bake, broil, boil, or steam foods instead of frying.
- Use nonstick cooking spray or a small amount of oil spread with a paper towel when cooking.
- Experiment with reducing oil in recipes. This may not affect the final product.
- Remove all visible fat and skin from meat and poultry.
- Use nonfat milk, 2% cottage cheese, and skim milk cheeses.
- Use egg whites or egg substitutes rather than whole eggs.
- Limit use of butter, mayonnaise, salad dressings, gravy, and sauces.
- Use corn or safflower oils and margarine.
- Use low-fat or "filled" cheeses in place of whole milk cheeses.

Healthy Choice: Reduce salt in your diet

- Never season food before tasting it. Take the salt shaker off the table.
- Reduce the amount of salt and high sodium ingredients in recipes. Each teaspoon of salt eliminated reduces sodium by 2132 milligrams.
- Use fewer canned, convenience, or highly processed foods.
- Watch for hidden sources of salt in meat tenderizer, soy, steak and Worcestershire sauces, relishes, pickles, olives, mustard, and catsup.
- Use less lunch meat, natural and process cheese, smoked fish and meats, sausage, bacon, hot dogs, and ham.

Substituting Herbs and Spices for Salt

Meat—Fish—Poultry

Beef:	Bay leaf, dry mustard powder, green pepper, marjoram
Chicken:	Green pepper, lemon juice, marjoram, fresh mushrooms, paprika, parsley, poultry seasoning, sage, thyme
Fish:	Bay leaf, curry powder, dry mustard powder, green pepper, lemon juice, marjoram, fresh mushrooms, paprika
Lamb:	Curry powder, garlic, mint, mint jelly, pineapple, rosemary
Pork:	Apple, applesauce, garlic, onion, sage
Veal:	Apricot, bay leaf, curry powder, ginger, marjoram, oregano

Vegetables

Asparagus:	Garlic, lemon juice, onion, vinegar
Corn:	Green pepper, pimento, fresh tomato
Cucumbers:	Chives, dill, garlic, vinegar
Green Beans:	Dill, lemon juice, marjoram, nutmeg, pimento
Greens:	Onion, pepper, vinegar
Peas:	Green pepper, mint, fresh mushrooms, onion, parsley
Potatoes:	Green pepper, mace, onion, paprika, parsley
Rice:	Chives, green pepper, onion, pimento, saffron
Squash:	Brown sugar, cinnamon, ginger, mace, nutmeg, onion
Tomatoes:	Basil, marjoram, onion, oregano
Soups:	A pinch of dry mustard powder in bean soup; allspice, a small amount of vinegar or a dash of sugar in vegetable soup; peppercorns in skim-milk chowders; bay leaf and parsley in pea soup

ALL-PURPOSE SEASONING I

2 teaspoons dried marjoram, crushed
2 teaspoons ground coriander seed
2 teaspoons paprika

½ teaspoon dry mustard
⅛ teaspoon garlic powder

Combine all ingredients. Store in tightly covered container in cool, dry place.

1 teaspoon = .68 mg. sodium

ALL-PURPOSE SEASONING II

1 tablespoon garlic powder
1 tablespoon dry mustard
1 tablespoon paprika

1½ teaspoons white pepper
1 teaspoon dried whole basil
½ teaspoon ground thyme

Combine all ingredients and mix well. Store in airtight container.

1 teaspoon = 1 mg. sodium

Healthy Choice: Increase fiber in your diet

- Choose fresh fruits rather than fruit juices.
- Eat high-fiber fruits such as apples, oranges, melons, berries, bananas, pears, and plums.
- Eat plenty of raw vegetables.
- Choose high-fiber vegetables such as beans, peas, parsnips, potatoes, broccoli, squash, carrots, tomatoes, green peppers, corn, asparagus, and pumpkin.
- Eat fruits and vegetables unpeeled when possible.
- Include unprocessed grain products in your daily diet such as whole-grain breads, bran cereal, brown rice, popcorn, grits, and oats.

Healthy Choice: Eat your vegetables

- Cook vegetables unpeeled and without removing leaves whenever possible.
- Use as little water as possible to bake, microwave, stir-fry, or steam.
- Cook vegetables in the shortest possible time to retain nutritional content.
- Keep vegetables whole as long as possible when cooking.
- Serve cooked vegetables and fruits as soon as possible after cooking.
- Eat deeply colored vegetables for more nutritional value.

Healthy Choice: Drink plenty of water

- Drink water *before* you are thirsty.
- Cold water is absorbed more rapidly than warm water.
- Drink 2 cups of water for each pound of weight lost during physical activity or in very hot weather.
- Fluids containing sugar are absorbed more slowly than water.
- Drink at least 8 glasses of water a day.

Breakfast Pizza

1 can refrigerator crescent rolls
½ lb. sausage
2 c. shredded Cheddar cheese
4 eggs, beaten
¾ c. milk
¼ tsp. oregano, pepper
½ tsp. salt

Fit crescent rolls into 9x9-inch baking pan, covering bottom and sides; press edges to seal. Brown sausage in skillet, stirring until crumbly; drain. Layer sausage and cheese in prepared pan. Combine remaining ingredients in bowl; mix well. Pour over layers. Bake at 400 degrees for 30 minutes. Yield: 6 servings.
Note: Nutritional information does not include refrigerator crescent rolls.

Approx per serving: Cal 413, Prot 18.4 gr,
T Fat 36.2 gr, Chol 233.5 mg, Carbo 2.6 gr,
Sod 777.1 mg, Potas 170.9 mg.

Sandra Whaley
North Whitfield Mid. Sch., Dalton, GA

Ham and Cottage Cheese Puffs

½ c. margarine
1 c. boiling water
1 c. flour
¼ tsp. salt
4 eggs
1 8½-oz. can crushed pineapple,
*   drained*
⅔ c. shredded carrots
½ c. small-curd cottage cheese, drained
½ c. unflavored yogurt
2 4½-oz. cans ham, drained

Melt margarine in boiling water in saucepan. Stir in flour and salt vigorously. Cook until mixture forms ball, stirring constantly. Remove from heat. Add eggs 1 at a time, beating until smooth. Drop by level tablespoonfuls into muffin cups. Bake at 400 degrees for 25 minutes. Cool on wire racks. Cut off top third of each puff; remove soft centers. Combine remaining ingredients in bowl. Spoon into puffs. Yield: 8 servings.

Approx per serving: Cal 366, Prot 12.1 gr,
T Fat 26.1 gr, Chol 151.3 mg, Carbo 20.4 gr,
Sod 679.5 mg, Potas 216.8 mg.

Linda Peterson
Cleveland, OH

Special Hash

4 c. chopped peeled potatoes
¼ c. oil
1 c. chopped onion
½ c. chopped green pepper
2 cloves of garlic, minced
1 lb. cooked corned beef, chopped
3 tbsp. Dijon mustard
¼ tsp. salt
¼ tsp. hot pepper sauce
4 to 6 poached eggs

Sauté potatoes in hot oil in large skillet for 10 to 15 minutes or until brown and crisp. Add onion, green pepper and garlic. Sauté until vegetables are tender. Stir in corned beef, mustard, salt and pepper sauce. Cook until heated through. Spoon onto serving plates. Top with poached egg. Yield: 4 to 6 servings.

Photograph for this recipe on page 9.

Zesty Egg-Tomato Cups

4 tomato cups
¼ tsp. salt
½ c. chopped scallions
½ c. chopped ham
1 tsp. margarine
4 eggs, beaten
1 tbsp. milk
¼ tsp. salt
Pepper to taste

Sprinkle tomato cups with ¼ teaspoon salt; invert to drain. Sauté scallions and ham in margarine in skillet. Beat eggs with milk, ¼ teaspoon salt and pepper. Pour into skillet. Cook until firm but not dry, stirring constantly. Spoon into tomato cups. Serve on lettuce-lined plates. Yield: 4 servings.

Approx per serving: Cal 175, Prot 12.0 gr,
T Fat 11.0 gr, Chol 269.0 mg, Carbo 7.0 gr,
Sod 476.0 mg, Potas 439.0 mg.

Kerri Rendelman
Newberry, NH

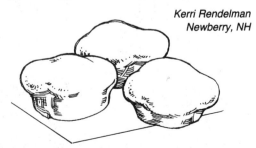

Honey-Wheat Germ Biscuits

4 c. flour
1 c. wheat germ
2 tbsp. baking powder
1¼ tsp. salt
1 c. butter
⅔ c. milk
½ c. honey

Mix flour, wheat germ, baking powder and salt in bowl. Cut in butter until crumbly. Add mixture of milk and honey; stir with fork until mixture forms ball. Roll ¾ inch thick on lightly floured surface. Cut with 3-inch biscuit cutter. Place 2 inches apart on ungreased baking sheet. Bake at 425 degrees for 10 to 12 minutes or until golden brown. Yield: 30 biscuits.

Approx per biscuit: Cal 149, Prot 3.0 gr,
T Fat 6.9 gr, Chol 19.7 mg, Carbo 19.4 gr,
Sod 232.7 mg, Potas 59.5 mg.

Deidra Roberts
Crystal Mid.-H.S., Carson City, MI

Yogurt Muffins

½ c. butter
1 c. whole wheat flour
¾ c. all-purpose flour
⅓ c. sugar
¼ tsp. salt
½ tsp. soda
1 egg, beaten
1 8-oz. carton flavored yogurt

Cut butter into dry ingredients in bowl until crumbly. Stir in egg and yogurt just until moistened. Spoon into greased muffin cups. Bake at 375 degrees for 20 minutes or until golden brown. Yield: 12 muffins.

Approx per muffin: Cal 170, Prot 3.4 gr,
T Fat 9.1 gr, Chol 46.4 mg, Carbo 19.6 gr,
Sod 187.3 mg, Potas 79.1 mg.

Amy Albitre
Ventura, CA

Buckwheat Cakes

1 pkg. dry yeast
1 tsp. sugar
¼ c. lukewarm water
2 c. buckwheat flour
1 c. graham flour
1¼ c. lukewarm water
¾ tsp. soda
1 tbsp. boiling water
2 tbsp. molasses
2 tbsp. oil
½ tsp. salt
½ c. buttermilk

Dissolve yeast and sugar in ¼ cup lukewarm water. Combine buckwheat and graham flour with 1¼ cups lukewarm water in bowl. Stir in yeast mixture. Let rise until light and bubbly. Stir in soda dissolved in 1 tablespoon boiling water and remaining ingredients. Bake on hot griddle until brown on both sides. Yield: 4 servings.

Approx per serving: Cal 362, Prot 8.9 gr,
T Fat 8.0 gr, Chol 6.0 mg, Carbo 66.7 gr,
Sod 463.0 mg, Potas 390.0 mg.

Janet H. Conrad
Pendleton, WV

Energy Pancakes

1¼ c. whole wheat flour
1 tbsp. baking powder
1 egg
2½ c. milk
2 tbsp. honey
2 tbsp. oil
¼ tsp. vanilla extract
¾ c. oats
¼ c. bran
¾ c. raisins
¼ c. each sunflower seed, sesame seed
¼ c. wheat germ
1 apple, shredded

Sift whole wheat flour and baking powder into bowl. Beat egg, milk and honey in mixer bowl until smooth. Add sifted ingredients alternately with oil and vanilla, mixing well after each addition. Stir in remaining ingredients. Chill overnight to 1 week. Spoon batter onto hot griddle. Bake until light brown on both sides. Serve with applesauce, fresh fruit or hot honey. Yield: 20 pancakes.

Approx per pancake: Cal 129, Prot 4.2 gr,
T Fat 5.2 gr, Chol 16.9 mg, Carbo 18.5 gr,
Sod 75.8 mg, Potas 186.4 mg.

Judy Christianson
Brillion H.S., Brillion, WI

Crunchy Breakfast Bars

> 2 c. oats
> ⅔ c. coconut
> ½ c. chopped pecans
> ⅓ c. wheat germ
> ¼ c. packed brown sugar
> ¼ c. margarine, softened
> 2 tbsp. honey

Mix oats, coconut, pecans and wheat germ in bowl. Blend brown sugar, margarine and honey in small bowl. Add to oats mixture; mix well. Press into well-greased 9x13-inch baking pan. Bake at 325 degrees for 15 to 20 minutes. Cool. Cut into bars. Yield: 24 bars.

Approx per bar: Cal 89, Prot 1.7 gr,
T Fat 5.0 gr, Chol 0.0 mg, Carbo 10.1 gr,
Sod 29.2 mg, Potas 67.7 mg.

Carla Seippel
Fort Osage Jr. H.S., Independence, MO

Granola

> 5 c. oats
> 1 c. coconut
> 1 c. sunflower seed
> 1 c. sesame seed
> 1 c. wheat germ
> 1 c. soy flour
> 1 c. honey
> 1 c. safflower oil
> 1 c. raisins

Combine first 6 ingredients in bowl. Add honey and oil; mix well. Place in ungreased baking pan. Bake at 300 degrees for 30 minutes or until lightly browned, stirring every 5 to 8 minutes. Combine with raisins in bowl. Cool. Store in airtight container. Yield: 10 cups.

Approx per cup: Cal 537, Prot 12.7 gr,
T Fat 36.1 gr, Chol 0.0 mg, Carbo 47.6 gr,
Sod 33.4 mg, Potas 586.8 mg.

Linda Darlington
Slavson Sch., Ann Arbor, MI

Blender Protein Breakfast

> 1 frozen banana, sliced
> 1 banana, sliced
> 1 c. milk
> 1 egg
> 3 tbsp. (heaping) protein powder
> 3 tbsp. (heaping) carob powder
> 3 tbsp. (heaping) wheat germ
> 1 c. ice cubes

Place bananas, milk, egg, protein powder, carob powder and wheat germ in blender container. Process until smooth. Add ice cubes gradually, processing until thickened. Pour into glasses. Yield: 3 servings.
Note: Nutritional information does not include protein powder.

Approx per serving: Cal 172, Prot 7.9 gr,
T Fat 5.7 gr, Chol 95.7 mg, Carbo 27.7 gr,
Sod 61.7 mg, Potas 428.5 mg.

Pat Foley
Tempe H.S., Tempe, AZ

Fruit Juicy Eggnog

> 2 eggs
> 2 tbsp. honey
> 1 c. apricot nectar
> ⅔ c. orange juice
> 1 tbsp. lemon juice
> ½ c. nonfat dry milk powder
> 1 c. frozen sweetened strawberries,
>   partially thawed

Beat eggs in mixer bowl until thick and lemon-colored. Add honey gradually, beating constantly; set aside. Combine apricot nectar, orange juice, lemon juice and dry milk powder in blender container. Process until smooth. Add strawberries. Process until smooth. Add egg mixture gradually, processing constantly. Pour into covered container. Chill in refrigerator. Pour into glasses. Garnish with mint leaf, strawberry slice and lemon twist. Yield: 4 servings.

Approx per serving: Cal 252, Prot 9.5 gr,
T Fat 3.3 gr, Chol 129.7 mg, Carbo 48.2 gr,
Sod 112.0 mg, Potas 554.0 mg.

Brian Jones
Torrance, NM

HEALTHY LUNCHES

Herbed Buttermilk Soup

2 cans tomato soup
2 soup cans buttermilk
2 tbsp. chopped fresh basil
½ tsp. sugar

Combine all ingredients with salt and pepper to taste in bowl; mix well. Chill in refrigerator for several hours to several days. Ladle into mugs or soup bowls to serve. Yield: 5 servings.

Approx per serving: Cal 134, Prot 6.4 gr,
T Fat 2.7 gr, Chol 2.4 mg, Carbo 22.1 gr,
Sod 1125.5 mg, Potas 400.9 mg.

Ella Jo Adams
Allen H.S., Allen, TX

Ground Beef and Oatmeal Soup

1 lb. ground beef
2 tbsp. oil
1 c. thinly sliced onion
1 c. sliced celery
3 c. water
1 can tomato soup
1 16-oz. can tomatoes
2½ tsp. salt
1 c. oats

Brown ground beef in oil in large saucepan, stirring until crumbly; drain. Add onion and celery. Cook, covered, until vegetables are tender; drain. Add water, soup, tomatoes and salt; mix well. Simmer, covered, for 10 minutes, stirring frequently. Stir in oats. Simmer just until thickened to desired consistency. Serve immediately. Yield: 6 servings.

Approx per serving: Cal 314, Prot 17.2 gr,
T Fat 17.8 gr, Chol 51.1 mg, Carbo 22.0 gr,
Sod 1449.6 mg, Potas 566.6 mg.

Leslie K. Donnell
Talawanda H.S., Oxford, OH

Cream of Potato Soup

2 tbsp. margarine
2 tbsp. flour
¼ tsp. salt
2 c. milk
2 c. frozen hashed brown potatoes
¼ c. water

⅛ tsp. parsley flakes
⅛ tsp. onion powder
⅛ tsp. celery salt
⅛ tsp. white pepper

Melt margarine in saucepan. Stir in flour and salt. Add milk gradually. Cook until bubbly and thickened, stirring constantly. Stir in remaining ingredients. Heat to serving temperature. Ladle into soup bowls. Yield: 2 servings.

Approx per serving: Cal 951, Prot 20.3 gr,
T Fat 20.1 gr, Chol 34.2 mg, Carbo 175.8 gr,
Sod 734.6 mg, Potas 1904.0 mg.

Nancy Doughty
Newcastle H.S., Newcastle, OK

Neptune Chef Salad

1 sm. head lettuce, shredded
1 c. sliced celery
1 c. shredded carrots
1 med. cucumber, sliced
1 sm. onion, chopped
1 c. sliced radishes
1 8-oz. package frozen peas, thawed
2 6-oz. cans water-pack chunk light tuna
1 c. shredded Cheddar cheese
1 8-oz. bottle of Thousand Island dressing

Place lettuce on salad plates. Layer celery, carrots, cucumber, onion, radishes, peas, tuna and cheese over lettuce. Drizzle with dressing. Yield: 4 servings.

Approx per serving: Cal 661, Prot 40.2 gr,
T Fat 42.5 gr, Chol 117.9 mg, Carbo 32.7 gr,
Sod 1617.0 mg, Potas 1136.6 mg.

Deborah C. Carter
Floyd Co. H.S., Floyd, VA

Tuna Riviera Platter

2 7-oz. cans tuna in vegetable oil
½ c. vinegar
¼ c. lemon juice
½ tsp. Tabasco sauce
2 tsp. dry mustard
¼ tsp. salt
1 tsp. sugar
4 c. assorted bite-sized fresh vegetables
Lettuce leaves

Drain oil from tuna into bowl, reserving tuna in cans. Add vinegar, lemon juice, Tabasco, dry mustard, salt and sugar to oil; mix well. Spoon 2 tablespoons marinade over tuna in each can. Place vegetables in remaining marinade. Chill vegetables and tuna for 1 hour or longer. Place lettuce leaves in center of serving platter. Mound drained tuna on lettuce. Arrange drained vegetables around tuna. Yield: 4 servings.

Photograph for this recipe on page 13.

Pasta Salad

2 c. snow peas
2 c. broccoli flowerets
1   16-oz. package cheese-stuffed
      tortelini
2½ c. cherry tomato halves
2 c. sliced fresh mushrooms
1   8-oz. can pitted ripe olives
1 tbsp. Parmesan cheese

Place snow peas in boiling water to cover in saucepan. Cook for 1 minute; drain. Place broccoli in boiling water to cover in saucepan. Cook for 1 minute; drain. Cook tortelini according to package directions; drain and cool slightly. Combine peas, broccoli, tomatoes, mushrooms, olives, pasta and Parmesan cheese in bowl. Add desired amount of Pasta Salad Dressing; mix well. Chill for several hours. Garnish with additional Parmesan cheese. Yield: 10 servings. Note: Nutritional information does not include the tortelini.

*Approx per serving: Cal 68, Prot 3.5 gr,
T Fat 3.2 gr, Chol 0.7 mg, Carbo 7.6 gr,
Sod 150.3 mg, Potas 267.7 mg.*

Pasta Salad Dressing

⅓ c. red wine vinegar
⅓ c. each vegetable oil, olive oil
½ c. sliced green onions
2 tbsp. chopped fresh parsley
2 cloves of garlic, minced
1½ tsp. Dijon mustard
½ tsp. sugar
2 tsp. basil
1 tsp. dillweed
½ tsp. each oregano, pepper
1 tsp. salt

Combine vinegar, vegetable oil and olive oil in jar. Add remaining ingredients; cover. Shake until well-mixed. Yield: 20 tablespoons.

*Approx per tablespoon: Cal 67, Prot 0.1 gr,
T Fat 7.2 gr, Chol 0.0 mg, Carbo 0.7 gr,
Sod 111.7 mg, Potas 14.2 mg.*

*Elsie S. Hilton
New Brockton Sch., New Brockton, AL*

Steak Salad with Mustard Dressing

1 lb. green beans, snapped
1 lb. 1-inch thick top round steak
⅓ c. Dijon mustard
1 tbsp. cider vinegar
1 tbsp. oil
¼ tsp. cracked pepper
½ tsp. salt
2 c. chopped lettuce
4 hard-boiled eggs

Cook green beans in a small amount of water in saucepan until tender-crisp; drain. Add salt to taste. Place steak on rack in broiler pan. Broil to desired degree of doneness. Slice into thin strips. Combine mustard, vinegar, oil, pepper and salt in bowl; mix well. Add green beans and steak; toss to coat well. Chill until serving time. Toss with chopped lettuce just before serving. Spoon into lettuce-lined serving dish. Slice 3 eggs. Chop remaining egg. Arrange over top of salad. Yield: 6 servings.

*Approx per serving: Cal 241, Prot 21.1 gr,
T Fat 14.5 gr, Chol 217.7 mg, Carbo 6.6 gr,
Sod 317.7 mg, Potas 434.8 mg.*

*Trudy K. Miller
Bishop Carroll H.S., Wichita, KS*

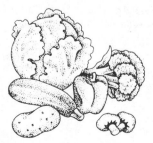

Tuna-Yogurt Salad

2 hard-boiled eggs
½ c. low-fat yogurt
½ tsp. mustard
¼ tsp. pepper
½ tsp. Old Bay spice
1 7-oz. can water-pack tuna
½ c. sliced apple
½ c. chopped celery
¼ c. chopped green onions
1 tbsp. minced onion
1 tbsp. chopped parsley

Mash egg yolks in bowl. Combine with yogurt, mustard and spices. Add chopped egg whites and remaining ingredients; mix well. Chill for 1 hour. Garnish with parsley sprigs. Serve with crackers, bagels or toast. Yield: 6 servings.

Approx per serving: Cal 114, Prot 12.3 gr,
T Fat 2.7 gr, Chol 105.1 mg, Carbo 10.0 gr,
Sod 320.0 mg, Potas 339.6 mg.

Martha Campbell
Jonesborough, TN

Broiled Deviled Hamburger Patties

1 lb. lean ground beef
2 tsp. minced onion
2 tsp. horseradish
¼ c. catsup
1½ tsp. prepared mustard
1½ tsp. Worcestershire sauce
1 tsp. salt
Pinch of pepper

Combine all ingredients in bowl; mix well. Shape into patties. Place on broiler rack. Broil 3 inches from heat source for 3 minutes on each side. Yield: 4 servings.

Approx per serving: Cal 207, Prot 23.8 gr,
T Fat 9.8 gr, Chol 79.9 mg, Carbo 4.7 gr,
Sod 790.7 mg, Potas 342.7 mg.

Dollie Edwards
Claxton H.S., Claxton, GA

Good Witch

12 flour tortillas
1 c. each chopped broccoli, onion
1 c. each shredded carrots, cabbage
1 c. alfalfa sprouts
1 c. sliced avocado
2 tbsp. chopped dill pickle
¼ c. salad dressing

Wrap tortillas in damp cloth. Warm in oven using package directions. Combine broccoli, onion, carrots and cabbage in glass bowl. Microwave until tender-crisp. Add alfalfa sprouts, avocado and dill pickle; mix gently. Spread tortillas with salad dressing. Spoon vegetable filling into tortillas. Roll tightly to enclose filling. Eat like sandwich. Yield: 24 servings.
Note: May add barbecue sauce to sandwich filling if desired.
Nutritional information does not include alfalfa sprouts.

Approx per serving: Cal 78, Prot 2.0 gr,
T Fat 3.0 gr, Chol 1.2 mg, Carbo 11.4 gr,
Sod 28.0 mg, Potas 112.1 mg.

Carol Curtis
Newburg, NC

Turkey-Vegetable Pockets

1½ c. cooked turkey strips
½ c. reduced-calorie mayonnaise
½ c. shredded carrot
½ c. chopped cucumber
1 sm. tomato, chopped
½ tsp. basil
2 pita bread rounds
4 lettuce leaves

Combine turkey, mayonnaise, carrot, cucumber, tomato and basil in bowl; mix gently. Chill in refrigerator. Cut pita bread rounds into halves. Fill with turkey mixture and lettuce leaf.
Yield: 4 servings.
Note: Nutritional information does not include reduced-calorie mayonnaise or pita bread.

Approx per serving: Cal 117, Prot 17.3 gr,
T Fat 3.3 gr, Chol 46.7 mg, Carbo 3.9 gr,
Sod 78.5 mg, Potas 378.5 mg.

Sybil B. Murphy
Northwood H.S., Pittsboro, NC

Homemade Pita Bread

4 c. flour
1½ tsp. salt
1 tsp. sugar
1 pkg. dry yeast
1¾ c. water
2 tbsp. peanut oil
1 c. flour

Mix 4 cups flour, salt, sugar and yeast in bowl. Combine water and peanut oil in small saucepan. Heat to 125 to 130 degrees. Stir into flour mixture. Add enough remaining 1 cup flour to make soft dough. Knead on floured surface for 8 to 10 minutes or until smooth and elastic. Let rest, covered, for 10 minutes. Divide into 12 portions. Roll each portion into 5-inch circle on floured surface. Place on wire racks on baking sheets. Bake at 500 degrees for 5 minutes. Yield: 12 pitas.

Approx per pita: Cal 215, Prot 5.7 gr, T Fat 2.8 gr, Chol 0.0 mg, Carbo 40.9 gr, Sod 267.8 mg, Potas 61.2 mg.

Tamara Friesen
Jardine Jr. H.S., Wichita, KS

Mock Barbecued Pork

4 oz. water-pack tuna
1   2-oz. can mushrooms
¼ green pepper, chopped
⅓ c. tomato paste
1 tbsp. mustard
1 tbsp. wine vinegar
1 tbsp. Worcestershire sauce
1 tbsp. onion flakes
1 tsp. chili powder

Combine all ingredients in small saucepan. Season with salt and pepper to taste. Simmer for 10 minutes. Serve over toast, English muffins or rice. Yield: 1 serving.

Approx per serving: Cal 264, Prot 39.0 gr, T Fat 1.9 gr, Chol 77.7 mg, Carbo 22.9 gr, Sod 1354.6 mg, Potas 1299.5 mg.

Sharon Ledgerwood
Moore H.S., Moore, OK

Shrimp Ratatouille

½ c. chopped onion
¼ c. olive oil
1 c. thinly sliced zucchini
2 c. cubed peeled eggplant
1 sm. sweet red pepper, sliced
1 sm. green pepper, sliced
2¼ c. peeled shrimp
1 c. sliced fresh mushrooms
1 c. canned tomatoes
1 tsp. basil
1½ tsp. garlic salt
¼ tsp. pepper
1 tbsp. chopped parsley

Sauté onion in olive oil in skillet. Add next 4 vegetables. Stir-fry for 5 minutes or until tender-crisp. Add shrimp. Stir-fry for 2 minutes. Add remaining ingredients. Cook, covered, for 5 minutes or until shrimp are tender. Serve with rice. Yield: 6 servings.

Approx per serving: Cal 229, Prot 25.0 gr, T Fat 10.4 gr, Chol 141.7 mg, Carbo 8.9 gr, Sod 725.0 mg, Potas 473.6 mg.

Mary Muller
Chesterfield, VA

Fruit Kabobs

1 c. grapefruit juice
½ c. honey
2 tbsp. Kirsch
1 tbsp. finely chopped fresh mint
6 canned peach halves, cut in half
4 bananas, cut into 2-in. pieces
2 apples, cut into wedges
1 fresh pineapple, chopped
2 grapefruit, sectioned

Combine first 4 ingredients in bowl; mix well. Arrange fruit in shallow bowl. Pour honey mixture over top. Let stand at room temperature for 1 hour. Thread fruit onto skewers. Place in broiler pan. Broil for 5 minutes, brushing frequently with marinade. Yield: 6 servings.

Approx per serving: Cal 370, Prot 2.1 gr, T Fat 0.9 gr, Chol 0.0 mg, Carbo 95.1 gr, Sod 7.4 mg, Potas 742.9 mg.

Socorro Perez
Fort Hancock, TX

Banana Praline Sundaes

3 tbsp. butter
1/3 c. packed brown sugar
1/4 c. golden raisins
2 med. bananas, sliced
1 pt. vanilla ice cream, softened
1/3 c. chopped pecans

Melt butter and brown sugar in skillet over low heat, stirring frequently. Add raisins and bananas. Cook for 5 minutes, stirring gently. Spoon ice cream into 4 serving dishes. Top with fruit mixture and pecans. Yield: 4 servings.

Approx per serving: Cal 442, Prot 3.6 gr, T Fat 27.6 gr, Chol 68.8 mg, Carbo 49.9 gr, Sod 137.9 mg, Potas 439.2 mg.

Kenny Singer
Huntingdon, WV

Frosty Drink

1 banana, mashed
1/2 c. each orange juice, cold milk
2 c. vanilla ice milk

Combine banana, orange juice, milk and 1 cup ice milk in blender container. Process until smooth. Pour into tall glasses. Scoop remaining ice milk into glasses. Yield: 2 servings.

Approx per serving: Cal 307, Prot 9.4 gr, T Fat 9.0 gr, Chol 34.7 mg, Carbo 49.3 gr, Sod 120.7 mg, Potas 643.4 mg.

Valene Wall
Wichita Northwest H.S., Wichita, KS

Hawaiian Milk Cooler

1 1/2 qt. vanilla ice cream
4 c. cold milk
3 c. pineapple juice
1/3 c. orange juice
1 tbsp. lemon juice

Scoop ice cream into punch bowl. Add remaining ingredients; mix gently. Serve immediately. Yield: 24 servings.

Approx per serving: Cal 110, Prot 3.1 gr, T Fat 5.0 gr, Chol 19.0 mg, Carbo 13.5 gr, Sod 41.6 mg, Potas 173.0 mg.

Delinda McCormick
Caldwell Co. H.S., Cadiz, KY

Monkey Milk

1 banana, frozen
1 c. low-fat milk
1 tsp. vanilla extract
1 pkt. Equal artificial sweetener
4 ice cubes

Cut banana into fourths. Combine with remaining ingredients in blender container. Process until smooth. Pour into glasses to serve.
Yield: 2 servings.
Note: Nutritional information does not include Equal artificial sweetener.

Approx per serving: Cal 92, Prot 4.5 gr, T Fat 1.4 gr, Chol 4.9 mg, Carbo 16.4 gr, Sod 61.5 mg, Potas 366.4 mg.

Sharon Ledgerwood
Moore H.S., Moore, OK

Orange Junior

1/2 c. milk
1/2 c. orange juice
2 c. orange sherbet

Combine milk, orange juice and 1 cup sherbet in blender container. Process for 10 seconds. Pour into glasses. Top with scoops of remaining sherbet. Yield: 3 servings.

Approx per serving: Cal 217, Prot 2.9 gr, T Fat 3.0 gr, Chol 5.7 mg, Carbo 45.9 gr, Sod 33.6 mg, Potas 169.5 mg.

Sandra Whaley
North Whitfield Mid. Sch., Dalton, GA

Orange Shake

2 c. vanilla ice cream
1/2 c. frozen orange juice concentrate
2 tbsp. honey
3 c. milk

Place ice cream in blender container. Add remaining ingredients; process until smooth. Pour into glasses. Yield: 6 servings.

Approx per serving: Cal 220, Prot 6.9 gr, T Fat 9.0 gr, Chol 34.8 mg, Carbo 28.7 gr, Sod 89.6 mg, Potas 425.7 mg.

Billie L. Perrin
Lafayette Co. H.S., Higginsville, MO

Carrot Dip For Broccoli

¾ c. cream-style cottage cheese
3 tbsp. mayonnaise
2 med. carrots, peeled, grated
3 sm. dill pickles, finely chopped
1 tbsp. caraway seed
½ tsp. salt
¼ tsp. pepper

Combine cottage cheese and mayonnaise in blender container. Process at high speed until smooth. Mix with remaining ingredients in bowl. Chill. Serve with broccoli flowerets for dipping. Yield: 48 tablespoons.

Approx per tablespoon: Cal 12, Prot 0.6 gr,
T Fat 0.9 gr, Chol 1.3 mg, Carbo 0.5 gr,
Sod 95.6 mg, Potas 21.9 mg.

Lois H. Webber
East Forsyth H.S., Kernersville, NC

Mexican Vegetable Dip

1 tbsp. each olive oil, vinegar
¼ tsp. garlic salt
1   20-oz. can black olives, chopped
2 med. green peppers, chopped
10 green onions, chopped
2 chili peppers, chopped
2 lg. tomatoes, chopped

Combine olive oil, vinegar, garlic salt and pepper to taste in bowl; mix well. Add chopped vegetables; mix well. Chill overnight; stir to mix. Serve with tortilla chips. Yield: 96 tablespoons.

Approx per tablespoon: Cal 12, Prot 0.1 gr,
T Fat 1.1 gr, Chol 0.0 mg, Carbo 0.5 gr,
Sod 43.0 mg, Potas 14.1 mg.

Judy Murray
Lake Mills H.S., Lake Mills, WI

Spinach Dip

20 oz. frozen chopped spinach
1 c. each sour cream, mayonnaise
1 env. Knorr dried vegetable soup mix
1 can water chestnuts, chopped

Cook spinach according to package directions; drain well. Combine with remaining ingredients in bowl; mix well. Chill until serving time. Serve with crackers. Yield: 64 tablespoons.
Note: Nutritional information does not include soup mix.

Approx per tablespoon: Cal 38, Prot 0.5 gr,
T Fat 3.6 gr, Chol 4.0 mg, Carbo 1.3 gr,
Sod 28.2 mg, Potas 42.7 mg.

Genia Wilson
Heritage H.S., Maryville, TN

Spinach Balls

20 oz. frozen chopped spinach
2 c. herb stuffing mix, crushed
1 c. Parmesan cheese
6 eggs, slightly beaten
¾ c. margarine, softened
¼ tsp. each thyme, pepper
½ tsp. seasoned salt

Cook spinach according to package directions; drain well. Mix with remaining ingredients in bowl. Shape into 1-inch balls. Place on greased baking sheet. Bake at 350 degrees for 10 minutes or until firm. Yield: 48 appetizers.

Approx per appetizer: Cal 59, Prot 2.5 gr,
T Fat 4.4 gr, Chol 34.1 mg, Carbo 2.7 gr,
Sod 128.6 mg, Potas 59.5 mg.

Elizabeth Wintersteen
Central Columbia H.S., Danville, KY

Cucumber Salad with Spicy Dressing

2 med. cucumbers, peeled
1 tbsp. white vinegar
1 tbsp. sugar
1 tsp. soy sauce
2 tsp. sesame oil
¼ tsp. Tabasco sauce
½ tsp. salt

Cut cucumbers into halves lengthwise. Scoop out seed. Combine seed with remaining ingredients in bowl. Chill cucumbers and seed mixture. Arrange cucumber halves on serving plate. Fill with seed mixture. Yield: 4 servings.

Approx per serving: Cal 56, Prot 1.4 gr,
T Fat 2.4 gr, Chol 0.0 mg, Carbo 8.5 gr,
Sod 385.7 mg, Potas 250.0 mg.

Kathy Thomas
Chickasha H.S., Chickasha, OK

Fresh Spinach Salad

8 oz. fresh mushrooms, sliced
1 c. Italian dressing
1 lb. fresh spinach, torn
6 slices crisp-fried bacon, crumbled
1/4 c. chopped green onion tops
1/3 c. grated Cheddar cheese
3 hard-boiled eggs, cut into wedges
1 tomato, cut into wedges

Mix mushrooms and Italian dressing in bowl. Chill for 2 to 3 hours. Combine with spinach in serving bowl. Add bacon, green onion, cheese and salt and pepper to taste; toss to mix well. Arrange egg and tomato wedges on salad. Garnish with paprika. Serve immediately. Yield: 6 servings.

Approx per serving: Cal 367, Prot 10.6 gr, T Fat 33.2 gr, Chol 139.2 mg, Carbo 9.5 gr, Sod 1047.6 mg, Potas 630.2 mg.

Marcia Ingram
Lexington H.S., Lexington, TX

Marinated Fresh Vegetables

Flowerets of 4 stalks broccoli
Flowerets of 1 sm. head cauliflower
8 lg. fresh mushrooms, sliced
3 stalks celery, chopped
1 med. green pepper, chopped
1 sm. onion, grated
2 tbsp. poppy seed
1 16-oz. bottle reduced-calorie Italian dressing

Combine vegetables and poppy seed in bowl. Add Italian dressing; mix gently. Chill for 3 hours or longer. Yield: 10 servings.

Approx per serving: Cal 71, Prot 4.6 gr, T Fat 2.6 gr, Chol 0.0 mg, Carbo 10.3 gr, Sod 413.7 mg, Potas 533.1 mg.

Edna O. Hutchins
Heritage H.S., Maryville, TN

Stir-Fry Beef and Spinach

1 1/2 lb. round steak
2 tbsp. soy sauce
1/4 tsp. Five-Spice Powder
2 tsp. cornstarch
1/4 c. water

1/4 tsp. instant beef bouillon
1 tbsp. oil
1 tsp. grated gingerroot
6 c. small fresh spinach leaves
1/2 c. sliced water chestnuts

Freeze beef partially. Slice into thin strips. Combine with soy sauce and Five-Spice Powder in bowl. Marinate for 15 minutes. Blend cornstarch and water in small bowl. Add bouillon; set aside. Heat oil in wok. Add gingerroot. Stir-fry for 30 seconds. Add beef mixture. Stir-fry for 2 to 3 minutes or until browned. Stir in cornstarch mixture. Cook until thickened, stirring constantly. Stir in spinach and water chestnuts. Cook for 1 to 2 minutes or until heated through. Yield: 6 servings.
Note: May substitute mixture of 1 teaspoon cinnamon, 1 teaspoon crushed anise, 1/4 teaspoon crushed fennel seed, 1/4 teaspoon pepper and 1/8 teaspoon cloves for Five-Spice Powder.

Approx per serving: Cal 265, Prot 25.0 gr, T Fat 14.6 gr, Chol 73.9 mg, Carbo 7.8 gr, Sod 575.5 mg, Potas 560.2 mg.

Wanda Watkins
Spartanburg H.S., Simpsonville, SC

Chicken Cacciatore

1 3 to 3 1/2-lb. chicken, cut up
1 sm. clove of garlic
2 tbsp. oil
1 tsp. oregano
1 16-oz. can stewed tomatoes
1/8 tsp. salt
Pepper to taste
1 1/2 c. sliced mushrooms

Sauté chicken with garlic in oil in skillet until lightly browned; remove garlic. Add remaining ingredients; mix well. Simmer, covered, for 30 minutes or until chicken is tender. Serve with spaghetti. Yield: 5 servings.

Approx per serving: Cal 170, Prot 18.3 gr, Fat 8.3 gr, Chol 61.4 mg, Carbo 5.0 gr, Sod 221.2 mg, Potas 480.4 mg.

Annette Wright
Philadelphia, PA

Golden Lasagna

 8 oz. lasagna noodles
 ½ c. chopped onion
 ½ c. green pepper
 1 tbsp. butter
 1 4-oz. can mushrooms, drained
 1 can cream of chicken soup
 ½ c. milk
 ½ tsp. basil
 1½ c. cottage cheese
 2 c. shredded Cheddar cheese
 ½ c. Parmesan cheese
 3 c. chopped cooked chicken

Cook noodles according to package directions. Sauté onion and green pepper in butter in saucepan. Add mushrooms, soup, milk and basil; mix well. Layer noodles, cheeses, chicken and sauce ½ at a time in 9 x 13-inch baking dish. Bake at 350 degrees for 45 minutes. Let stand for 10 minutes before serving. Yield: 8 servings.

Approx per serving: Cal 449, Prot 38.0 gr, T Fat 19.6 gr, Chol 308.2 mg, Carbo 27.5 gr, Sod 720.0 mg, Potas 408.8 mg.

Janet Ward
Paola H.S., Paola, KS

Stir-Fry Chicken and Vegetables

 3 tbsp. corn oil
 1 lb. boneless chicken, cut into
    bite-sized pieces
 5 tbsp. teriyaki sauce
 1 tsp. curry powder
 1 18-oz. package frozen
    oriental vegetables
 1 8-oz. can sliced bamboo shoots,
    drained
 1 green pepper, cut into thin strips
 1 c. diagonally sliced carrots

Heat oil in wok over high heat; reduce heat to medium. Add chicken. Stir-fry for 2 minutes. Sprinkle with teriyaki sauce and curry powder. Cook for 3 minutes longer, stirring frequently. Remove chicken with slotted spoon. Add vegetables. Stir-fry for 3 minutes. Add chicken. Stir-fry for 3 minutes longer. Serve over rice. Yield: 6 servings.

Approx per serving: Cal 214, Prot 14.7 gr, T Fat 8.3 gr, Chol 18.4 mg, Carbo 23.6 gr, Sod 661.6 mg, Potas 1019.7 mg.

Linda Finley
Harrison Central Ninth Gr. Sch., Gulfport, MS

Fish Florentine

 1 lb. flounder fillets
 1 med. onion, sliced into rings
 ¼ tsp. salt
 1 tsp. coarsely ground pepper
 ½ c. chicken broth
 2 tbsp. white wine
 ½ c. evaporated skim milk
 1 tbsp. flour
 ½ tsp. dillweed
 ¼ tsp. oregano
 1 10-oz. package frozen spinach,
    cooked, drained

Layer fillets and onion in 10-inch skillet. Sprinkle with salt and pepper. Add broth and wine. Bring to a boil; reduce heat. Simmer, covered, for 7 to 8 minutes or until fish flakes easily. Combine evaporated milk, flour, dillweed and oregano in bowl; mix well. Stir into skillet. Cook until thickened, stirring gently. Simmer for 1 to 2 minutes longer. Spoon spinach onto serving plate. Place fish on spinach. Spoon sauce over top. Yield: 4 servings.
Note: Nutritional information does not include evaporated skim milk.

Approx per serving: Cal 144, Prot 37.2 gr, T Fat 0.3 gr, Chol 56.7 mg, Carbo 8.0 gr, Sod 387.3 mg, Potas 989.4 mg.

Wanda Watkins
Spartanburg H.S., Simpsonville, SC

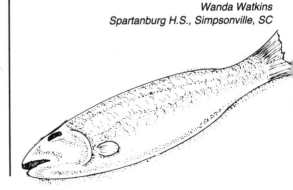

Herbed Baked Fish

½ c. butter
⅔ c. cracker crumbs
¼ c. Parmesan cheese
½ tsp. basil
½ tsp. oregano
¼ tsp. garlic powder
½ tsp. salt
1 lb. fish fillets

Melt butter in 9x13-inch baking dish. Combine cracker crumbs, Parmesan cheese, basil, oregano, garlic powder and salt in bowl; mix well. Roll fish in butter. Coat with crumb mixture. Arrange in remaining butter in dish. Bake at 350 degrees for 25 to 30 minutes or until fish flakes easily. Yield: 4 servings.

Approx per serving: Cal 459, Prot 25.3 gr, T Fat 35.6 gr, Chol 140.5 mg, Carbo 8.7 gr, Sod 789.1 mg, Potas 370.8 mg.

Kathy Thomas
Chickasha Mid. Sch., Chickasha, OK

Sesame Fish Fillets

1 tbsp. sesame seed
¼ c. teriyaki sauce
3 tbsp. orange juice
1 tbsp. Brandy
1 tsp. sesame oil
¼ tsp. ginger
4 fillets of sole or flounder
1 tbsp. slivered green onion

Heat sesame seed in small saucepan until golden, stirring frequently; set aside. Combine teriyaki sauce, orange juice, Brandy, sesame oil and ginger in bowl; mix well. Rinse fish fillets; pat dry. Fold fillets in half crosswise; place each on 12-inch square of heavy foil. Drizzle with teriyaki sauce mixture. Sprinkle with sesame seed and green onion. Fold foil to enclose fish, sealing well. Grill 5 inches from medium-hot coals for 10 to 12 minutes or until fish flakes easily. Yield: 4 servings.

Photograph for this recipe on page 19.

Skewered Scallops

1½ lb. sea scallops
1 lg. cantaloupe
3 tbsp. fresh lime juice
2 tbsp. minced fresh mint leaves
1½ tbsp. honey
¾ tsp. salt
2 tbsp. melted butter

Soak 12 wooden skewers in water. Rinse scallops and pat dry. Scoop cantaloupe into melon balls. Combine lime juice, mint, honey and salt in bowl; mix well. Add scallops and melon balls. Chill, covered, for 1 hour. Pat skewers dry. Drain scallops and melon balls. Thread onto skewers. Brush with butter. Grill 5 inches from medium-hot coals for 2 minutes; turn. Brush with butter. Grill for 3 minutes or until scallops are firm. Yield: 6 servings.

Photograph for this recipe on page 19.

Fantastic Veggie Pie

2 pkg. refrigerator crescent dinner rolls
16 oz. cream cheese, softened
1 env. ranch dressing mix
1 c. mayonnaise
½ c. each finely chopped mushrooms,
    broccoli, cauliflower, carrots,
    green pepper, and cucumber
1 c. grated Cheddar cheese

Unroll crescent roll dough. Place in 9x13-inch baking dish; press edges to seal. Bake at 400 degrees for 10 minutes. Cool. Blend cream cheese, ranch dressing mix and mayonnaise in bowl. Spread over crust. Combine vegetables in bowl. Sprinkle over cream cheese mixture. Sprinkle Cheddar cheese on top. Chill, covered, with waxed paper, overnight. Cut into squares. Yield: 8 servings.
Note: Nutritional information does not include crescent dinner rolls or ranch dressing mix.

Approx per serving: Cal 483, Prot 9.4 gr, T Fat 48.4 gr, Chol 96.5 mg, Carbo 4.8 gr, Sod 416.3 mg, Potas 208.9 mg.

Annmarie Milnamow
St. Edward H.S., Elgin, IL

Stir-Fry Carrots and Green Beans

½ lb. fresh green beans
½ lb. carrots
2 tbsp. oil
1 sm. onion, sliced
½ tsp. salt

Trim beans and snap into 1-inch pieces. Slice carrots ¼ inch thick. Combine beans and carrots with a small amount of water in saucepan. Steam, covered, for 5 minutes; drain. Heat oil in large skillet over high heat. Stir-fry onion for 3 minutes. Sprinkle with salt. Add carrots and beans. Stir-fry for 1 minute. Yield: 2 servings.

Approx per serving: Cal 220, Prot 4.0 gr,
T Fat 14.1 gr, Chol 0.0 mg, Carbo 22.7 gr,
Sod 598.4 mg, Potas 729.0 mg.

Kathleen Burchett
Area Supr. of H.E., State Dept. of Ed., Abingdon, VA

Spinach Casserole

1 10-oz. package frozen chopped
*   spinach, thawed, drained*
1 16-oz. carton cream-style
*   cottage cheese*
6 eggs, beaten
1 stick margarine, melted
2 c. grated American cheese
6 tbsp. flour

Combine all ingredients in bowl; mix well. Spoon into 1½-quart baking dish. Bake at 350 degrees for 1 hour. Yield: 8 servings.

Approx per serving: Cal 350, Prot 20.6 gr,
T Fat 25.4 gr, Chol 221.7 mg, Carbo 10.1 gr,
Sod 799.1 mg, Potas 300.1 mg.

Sue Smith
Bartlesville H.S., Bartlesville, OK

Garden Scramble Stir-Fry

1 tbsp. oil
1 lg. clove of garlic, minced
1 c. broccoli flowerets
1 c. cauliflowerets
2 tbsp. water
1 tbsp. oil
½ c. carrot chunks
½ c. green pepper slices
1 tbsp. water

Heat 1 tablespoon oil in wok over high heat. Add garlic. Stir-fry for several seconds. Add broccoli and cauliflower. Stir-fry for 1 minute. Add 2 table-spoons water. Cook, covered, for 3 minutes, stirring frequently. Remove vegetables with slotted spoon. Add remaining tablespoon oil to wok. Add carrots and green pepper. Stir-fry for 1 minute. Add 1 tablespoon water. Cook, covered, for 2 minutes or until tender-crisp, stirring frequently. Stir in broccoli mixture and salt and pepper to taste. Garnish with cashews. Yield: 4 servings.

Approx per serving: Cal 96, Prot 2.4 gr,
T Fat 7.1 gr, Chol 0.0 mg, Carbo 7.5 gr,
Sod 25.0 mg, Potas 348.6 mg.

Pat Vaughan
Fairfield H.S., Fairfield, IL

Stuffed Zucchini

3 med. zucchini
1 c. chopped fresh mushrooms
2 tbsp. butter
2 tbsp. flour
¼ tsp. oregano
2 tbsp. chopped pimento
1 c. shredded provolone cheese
¼ c. Parmesan cheese

Cook whole zucchini in boiling water in saucepan for 10 minutes or until tender; drain. Cut zucchini in half lengthwise. Scoop out and chop centers, leaving ¼-inch shells. Sauté mushrooms in butter in skillet for 3 minutes. Stir in flour and oregano; remove from heat. Add pimento, provolone cheese and chopped zucchini. Cook until heated through. Spoon into zucchini shells. Place on baking sheet. Sprinkle with Parmesan cheese. Broil for 3 to 5 minutes or until bubbly. Yield: 6 servings.

Approx per serving: Cal 149, Prot 8.0 gr,
T Fat 9.9 gr, Chol 33.2 mg, Carbo 8.2 gr,
Sod 164.7 mg, Potas 346.6 mg.

Evelyn Marvin
Fowler Sch., Fowler, CO

Dried Fruit Compote

1 8-oz. package dried pitted prunes
1 8-oz. package dried apricots
1 20-oz. can pineapple chunks
1 11-oz. can mandarin oranges
1 20-oz. can cherry pie filling
½ c. water

Layer prunes, apricots, pineapple with juice and oranges with juice in 3-quart baking dish. Place pie filling in bowl. Rinse pie filling can with water. Mix water with pie filling. Pour over fruit layers. Bake at 350 degrees for 1 hour. Serve hot with meats. Yield: 12 servings.

Approx per serving: Cal 202, Prot 1.7 gr,
T Fat 0.3 gr, Chol 0.0 mg, Carbo 51.7 gr,
Sod 7.4 mg, Potas 394.2 mg.

Loretta Jane Pesterfield
Heritage H.S., Maryville, TN

Whole Wheat Biscuits

2 c. whole wheat flour
1 tbsp. baking powder
1 tbsp. sugar
1 tsp. salt
¼ c. shortening
⅔ c. milk

Mix whole wheat flour, baking powder, sugar and salt in bowl. Cut in shortening until crumbly. Stir in milk. Shape lightly into ball on floured surface. Roll ½ inch thick. Cut with biscuit cutter. Place on greased baking sheet. Bake at 425 degrees for 10 minutes. Yield: 12 biscuits.

Approx per biscuit: Cal 122, Prot 3.1 gr,
T Fat 5.5 gr, Chol 1.9 mg, Carbo 16.1 gr,
Sod 267.2 mg, Potas 94.8 mg.

Betty Barnes
Beulah Sch., Valley, AL

Hearty Wheat Rolls

2 c. all-purpose flour
2 pkg. dry yeast
1 c. milk
¾ c. water
½ c. oil
½ c. sugar
1 tbsp. salt
2½ to 3 c. whole wheat flour

Mix all-purpose flour and yeast in mixer bowl. Combine milk, water, oil, sugar and salt in saucepan. Heat to 120 to 130 degrees. Pour into flour mixture. Beat at medium speed for 2 minutes. Add whole wheat flour 1 cup at a time, mixing well after each addition. Dough will be soft. Let rest for 12 to 15 minutes. Roll ½ inch thick on floured surface. Cut as desired. Place in greased pan. Let rise for 30 minutes. Bake at 400 degrees for 15 to 18 minutes or until brown. Yield: 18 rolls.
Note: Large rolls baked on baking sheet may be sliced and used as sandwich buns.

Approx per roll: Cal 203, Prot 4.9 gr,
T Fat 7.1 gr, Chol 1.9 mg, Carbo 31.3 gr,
Sod 363.4 mg, Potas 122.5 mg.

Katherine Anderson
East Central H.S., Lucedale, MS

Sugar-Free Apple Pie

1 tbsp. cornstarch
1 6-oz. can frozen apple juice
   concentrate, thawed
6 c. sliced peeled apples
1 recipe 2-crust pie pastry
1 tsp. cinnamon
½ tsp. nutmeg

Dissolve cornstarch in ¼ cup apple juice in small bowl. Combine with remaining apple juice in saucepan. Cook over medium heat until thickened, stirring constantly. Stir in apples, coating well. Spoon into pastry-lined 9-inch pie plate. Sprinkle with spices. Top with remaining pastry; trim edge and cut vents. Bake at 375 degrees for 50 to 60 minutes or until brown. Yield: 6 servings.
Note: May be used in diabetic diets.

Approx per serving: Cal 438, Prot 4.1 gr,
T Fat 20.8 gr, Chol 0.0 mg, Carbo 60.9 gr,
Sod 370.0 mg, Potas 297.2 mg.

Jane Bigler
Hillcrest Jr. H.S., Lenexa, KS

Yogurt-Fruit Sundaes

> 1   8-oz. carton plain yogurt
> 2 tbsp. honey
> ½ c. each sliced apples, bananas
> ½ c. orange sections
> ½ c. pineapple chunks
> ½ c. granola

Mix yogurt and honey in bowl. Place mixture of fruit in dessert glasses. Top each with yogurt mixture and 2 tablespoons granola. Garnish with fresh fruit slice. Yield: 4 servings.
Note: Nutritional information does not include granola.

Approx per serving: Cal 137, Prot 2.5 gr,
T Fat 2.3 gr, Chol 4.9 mg, Carbo 29.1 gr,
Sod 30.3 mg, Potas 285.7 mg.

Ruthie Andrews
Paul West Mid. Sch., East Point, GA

Hot Spicy Apple Juice

> 1 qt. unsweetened apple juice
> 1 c. water
> 2 whole cloves
> 2 whole allspice
> 1 stick of cinnamon

Combine apple juice and water in 8-cup electric percolator. Place spices in percolator basket. Perk until cycle is complete. Pour into cups. Yield: 6 servings.

Approx per serving: Cal 67, Prot 0.3 gr,
T Fat 0.3 gr, Chol 0.0 mg, Carbo 17.6 gr,
Sod 3.3 mg, Potas 126.9 mg.

Shirley Henkel
Statesville H.S., Statesville, NC

Fireside Sipper

> 1 stick of cinnamon
> 6 whole allspice
> 6 whole cloves
> 1 qt. unsweetened apple juice
> ½ to ¾ cup apricot nectar
> 2 to 4 tbsp. brown sugar

Tie spices in cheesecloth bag. Combine juices and brown sugar in medium saucepan. Bring to a simmer, stirring until brown sugar is dissolved.

Add spice bag. Simmer for 15 minutes. Remove spices. Pour into mugs. Garnish with apple wedge. Yield: 6 servings.

Approx per serving: Cal 119, Prot 0.4 gr,
T Fat 0.4 gr, Chol 0.0 mg, Carbo 31.0 gr,
Sod 6.0 mg, Potas 205.8 mg.

Sharie Mueller
Jefferson Co. North H.S., Oskaloosa, KS

Instant Russian Tea Mix

> 1 c. instant orange breakfast
>   drink mix
> 2 pkg. instant lemonade mix
> 1 c. instant tea
> 2½ c. sugar
> 2 tsp. cinnamon
> 1 tsp. cloves

Combine all ingredients in bowl; mix well. Store in airtight container. Dissolve 2 teaspoons mix in 1 cup hot water for each serving.

Betty K. Munsey
Bland H.S., Bland, VA

Fresh Lemon Freeze

> 3 pt. vanilla ice cream
> 1¼ c. fresh lemon juice
> 2½ c. crushed ice

Combine all ingredients in blender container. Process until thickened and smooth. Pour into glasses. Garnish with lemon slices. Yield: 10 servings.

Approx per serving: Cal 162, Prot 3.7 gr,
T Fat 8.5 gr, Chol 31.9 mg, Carbo 19.0 gr,
Sod 50.6 mg, Potas 187.3 mg.

Judy Meek
Marshall Jr. H.S., Wichita, KS

SALADS

Cranberry Salad

1 lg. package raspberry gelatin
1   6-oz. package cranberries
1 sm. orange, chopped
1 sm. apple, chopped
½ c. chopped celery
½ c. chopped pecans
1   8-oz. can crushed pineapple, drained
¼ c. sugar

Prepare gelatin in bowl according to package directions, using half the water. Chill until partially set. Place cranberries in blender container with a small amount of water. Process at medium speed for several seconds; drain. Repeat process with orange. Combine cranberries, orange, apple, celery, pecans, pineapple and sugar in bowl. Stir into partially set gelatin. Pour into 9x13-inch dish. Chill until firm. Yield: 10 servings.

Approx per serving: Cal 172, Prot 3.0 gr,
T Fat 4.1 gr, Chol 0.0 mg, Carbo 32.2 gr,
Sod 106.1 mg, Potas 286.3 mg.

Wynn Tillman
Charlotte, NC

Fast Frozen Salad

1 c. strawberry yogurt
¼ c. honey
4 oz. cream cheese, softened
1   7-oz. can crushed pineapple

Combine all ingredients in bowl; mix well. Spoon into paper-lined muffin cups. Freeze until firm. Remove from freezer 10 to 15 minutes before serving. Yield: 8 servings.

Approx per serving: Cal 127, Prot 6.4 gr,
T Fat 6.4 gr, Chol 18.2 mg, Carbo 16.6 gr,
Sod 50.7 mg, Potas 86.9 mg.

Michelle Presley
Hartsville, TN

Fruit Salad

1   16-oz. can pineapple chunks
1   16-oz. can fruit cocktail
1   16-oz. can peaches, chopped
1   3-oz. package strawberry
*   instant pudding mix*
2 bananas, sliced

1   16-oz. package frozen
*   whole strawberries*

Combine pineapple, fruit cocktail and peaches in bowl. Sprinkle with pudding. Stir in bananas and strawberries. Chill until serving time. Yield: 8 servings.

Approx per serving: Cal 287, Prot 1.8 gr,
T Fat 0.6 gr, Chol 0.0 mg, Carbo 74.0 gr,
Sod 70.4 mg, Potas 428.7 mg.

Beverly C. Goodman
Smyth Co. Voc. Sch., Marion, VA

Mandarin Orange Salad

1 head lettuce, torn
4 sm. red onions, sliced
1 c. chopped celery
1   11-oz. can mandarin oranges, drained
¼ c. white vinegar
⅔ c. oil
½ c. sugar
1 to 2 tbsp. poppy seed
1 tsp. mustard
1 tsp. salt

Combine lettuce, onions, celery and mandarin oranges in bowl. Chill until serving time. Combine remaining ingredients in jar; cover. Shake to mix well. Pour over salad at serving time; toss gently. Yield: 6 servings.

Approx per serving: Cal 336, Prot 2.2 gr,
T Fat 24.5 gr, Chol 0.0 mg, Carbo 30.5 gr,
Sod 405.8 mg, Potas 380.0 mg.

Carla Seippel
Fort Osage Jr. H.S., Independence, MO

Congealed Orange Salad

2 sm. packages orange gelatin
2 c. boiling water
6 oz. frozen orange juice concentrate
1   11-oz. can each mandarin oranges
*   and pineapple tidbits, drained*
3 bananas, sliced
1 c. pecan halves
2 c. dry-curd cottage cheese

Dissolve gelatin in boiling water in bowl. Cool. Stir in orange juice concentrate, oranges, pineapple, bananas and pecans. Pour into ring mold.

Chill until set. Unmold onto serving plate. Spoon cottage cheese into center. Yield: 10 servings.

Approx per serving: Cal 266, Prot 8.7 gr,
T Fat 8.7 gr, Chol 1.7 mg, Carbo 2.0 gr,
Sod 139.6 mg, Potas 441.7 mg.

Helen Love
Franklin, KY

Pretty and Good Fruit Salad

1   16-oz. can each reduced-calorie
    apricots, peaches, drained
1   11-oz. can mandarin oranges, drained
1   20-oz. can pineapple chunks
4 bananas, sliced
2 c. fresh strawberries
3 apples, sliced
1 sm. package sugar-free
    strawberry gelatin

Combine fruit in bowl; mix well. Sprinkle with gelatin. Chill for several hours. Serve with whipped topping if desired. Yield: 12 servings.
Note: Nutritional information does not include sugar-free gelatin.

Approx per serving: Cal 152, Prot 1.4 gr,
T Fat 0.6 gr, Chol 0.0 mg, Carbo 38.7 gr,
Sod 9.2 mg, Potas 350.0 mg.

Emma Ellen Bunyard
Jenks H.S., Jenks, OK

Raspberry Salad

2 sm. packages raspberry gelatin
2 c. hot water
1   16-oz. can cranberry sauce
1   20-oz. can crushed pineapple
½ c. chopped celery
1 c. chopped pecans

Dissolve gelatin in hot water in bowl. Add cranberry sauce; mix well. Stir in remaining ingredients. Pour into mold. Chill until set. Unmold onto serving plate. Yield: 12 servings.

Approx per serving: Cal 228, Prot 2.5 gr,
T Fat 7.2 gr, Chol 0.0 mg, Carbo 41.7 gr,
Sod 52.3 mg, Potas 170.9 mg.

Peggy S. Christian
Hampton H.S., Carrollton, VA

Ribbon Rainbow Gelatin Salad

5 sm. packages gelatin
    in assorted colors
5 c. boiling water
10 tbsp. cold water
10 tbsp. plain yogurt

Dissolve 1 package gelatin in 1 cup boiling water in bowl. Divide into 2 portions. Stir 2 tablespoons cold water into 1 portion. Pour into deep 8x8-inch pan. Chill until set. Stir 2 tablespoons yogurt into remaining portion. Pour over congealed layer. Chill until set. Repeat process 1 package at a time with remaining gelatin, chilling each layer until firmly set before adding the next layer. Chill until firm. Yield: 16 servings.
Note: May freeze cold water in 2 tablespoon portions to speed up chilling process.

Approx per serving: Cal 105, Prot 2.8 gr,
T Fat 0.3 gr, Chol 0.8 mg, Carbo 23.8 gr,
Sod 89.0 mg, Potas 68.5 mg.

Barbara Bird
Alma H.S., Alma, MI

Yogurt-Fruit Salad

1 c. each mandarin oranges, pineapple
    chunks, banana slices, and grapes
1 c. flaked coconut
1 c. unflavored or vanilla yogurt

Combine all ingredients in bowl; toss to coat. Chill for 30 minutes. Serve in lettuce cups. Yield: 5 servings.

Approx per serving: Cal 218, Prot 3.1 gr,
T Fat 7.1 gr, Chol 4.0 mg, Carbo 38.6 gr,
Sod 61.0 mg, Potas 437.0 mg.

Georgia Ann Jones
Harrison, WV

Creamy Fruit Dressing

¼ c. sour cream
3 tbsp. pineapple juice or mashed
    strawberries
1 c. mayonnaise
2 tbsp. chopped nuts

Combine all ingredients in bowl; mix well. Chill. Serve over fresh fruit salad. Yield: 1¼ cups.

Photograph for this recipe on page 27.

Poppy Seed Dressing

½ c. mayonnaise
2 tbsp. sugar
1 tbsp. lemon juice
1 tbsp. poppy seed

Combine all ingredients in bowl; mix well. Serve with fresh fruit salad. Yield: 12 tablespoons.

Approx per tablespoon: Cal 75, Prot 0.1 gr,
T Fat 7.5 gr, Chol 6.5 mg, Carbo 2.4 gr,
Sod 55.8 mg, Potas 5.0 mg.

Sandra Whaley
North Whitfield Mid. Sch., Dalton, GA

Almond-Chicken Salad

4 chicken breasts, skinned
1 tsp. each salt and pepper
1 tsp. lemon juice
1 c. salad dressing
½ c. roasted slivered almonds

Cook chicken breasts in water to cover in saucepan for 20 to 25 minutes or until tender. Drain and chop chicken. Combine with remaining ingredients in bowl; mix well. Chill, tightly covered, in refrigerator. Serve with fruit, cheese, pickles and crackers. Yield: 4 servings.

Approx per serving: Cal 534, Prot 37.3 gr,
T Fat 37.9 gr, Chol 103.5 mg, Carbo 11.7 gr,
Sod 935.9 mg, Potas 438.3 mg.

Sandra Whaley
North Whitfield Mid. Sch., Dalton, GA

Chicken Salad

½ c. melted butter, cooled
2 c. reduced-calorie mayonnaise
¼ c. minced parsley
½ tsp. minced garlic
Pinch each of marjoram, white pepper
4 chicken breasts, cooked, chopped
½ to 1 lb. green grapes
½ c. toasted almonds

Combine butter, mayonnaise, parsley, garlic, marjoram, white pepper and salt to taste in bowl. Add chicken, grapes and almonds; mix well. Chill until serving time. Serve on leaf lettuce with toast strips. Yield: 8 servings.
Note: Nutritional information does not include reduced-calorie mayonnaise.

Approx per serving: Cal 271, Prot 18.4 gr,
T Fat 17.4 gr, Chol 77.0 mg, Carbo 11.5 gr,
Sod 176.6 mg, Potas 386.9 mg.

Nora Sweat
West Hardin H.S., Elizabethtown, KY

Low-Cal Chicken Salad

1 3 to 4-lb. chicken
½ 6-oz. jar pickle relish
4 to 6 oz. Weight Watchers mayonnaise

Cook chicken in water to cover in saucepan until tender. Remove chicken. Cool for 10 to 15 minutes. Bone and chop chicken. Combine with remaining ingredients in bowl; mix well. Chill overnight. Yield: 10 servings.
Note: Nutritional information does not include Weight Watchers mayonnaise.

Approx per serving: Cal 296, Prot 23.8 gr,
T Fat 20.2 gr, Chol 122.6 mg, Carbo 3.1 gr,
Sod 114.0 mg, Potas 265.0 mg.

Juanita Boyce
Lawton H.S., Lawton, OK

Peachy Chicken Salad

2 c. chopped cooked chicken
¾ c. chopped celery
¾ c. seedless grapes
¾ c. chopped fresh peaches
½ c. each mayonnaise, sour cream
½ tsp. seasoned salt

Combine chicken, celery, grapes and peaches in bowl. Mix mayonnaise and sour cream in small bowl. Pour over salad; mix well. Add seasoned salt. Chill until serving time. Garnish with additional peach slices and parsley.
Yield: 6 servings.

Approx per serving: Cal 300, Prot 15.0 gr,
T Fat 23.1 gr, Chol 62.2 mg, Carbo 8.0 gr,
Sod 345.3 mg, Potas 307.2 mg.

James Gleaton
Lexington, KY

Turkey Salad

2 c. chopped cooked turkey breast
½ c. chopped celery
¼ c. chopped onion
¼ c. chopped sweet red pepper
1 c. low-fat yogurt
¼ tsp. curry powder

Combine turkey with celery, onion and red pepper in bowl; mix well. Add yogurt; mix well. Blend in curry powder. Chill for several hours. Spoon onto lettuce-lined serving plates. Garnish with paprika and sliced ripe olives. Yield: 4 servings.

Approx per serving: Cal 112, Prot 16.3 gr, T Fat 2.7 gr, Chol 37.2 mg, Carbo 4.9 gr, Sod 84.8 mg, Potas 343.3 mg.

Peggy Davis
Washington, D.C.

Zesty Low-Calorie Crab Salad

½ c. reduced-calorie mayonnaise
2 tbsp. chopped parsley
2 tbsp. minced chives
1 tsp. Worcestershire sauce
½ tsp. lemon juice
2   7-oz. cans king crab meat, drained
1 c. chopped cucumber
4 lettuce cups
4 hard-boiled eggs, cut into quarters

Combine mayonnaise, parsley, chives, Worcestershire sauce and lemon juice in bowl. Add crab meat and cucumber; toss to mix well. Chill, covered, for 1 hour or longer. Spoon into lettuce cups. Top with egg quarters. Yield: 4 servings. Note: Nutritional information does not include reduced-calorie mayonnaise.

Approx per serving: Cal 211, Prot 24.7 gr, T Fat 8.4 gr, Chol 343.7 mg, Carbo 9.7 gr, Sod 1001.4 mg, Potas 695.7 mg.

Sue P. Culpepper
Gulfport H.S., Gulfport, MS

Shrimp and Melon Salad

3 cantaloupes
2 c. chilled cooked shrimp
1 c. chopped celery
¼ c. mayonnaise
1 tbsp. lemon juice
¼ tsp. salt
⅛ tsp. pepper

Cut cantaloupes in half; discard seed. Combine shrimp with remaining ingredients in bowl; mix lightly. Place cantaloupe halves on lettuce-lined plates. Spoon shrimp mixture into cantaloupes. Yield: 6 servings.

Approx per serving: Cal 216, Prot 15.9 gr, T Fat 8.4 gr, Chol 41.6 mg, Carbo 21.5 gr, Sod 280.9 mg, Potas 809.3 mg.

Judy Myron
Charlottesville, VA

Vegetable-Shrimp Salad

1½ c. torn spinach
½ c. sliced mushrooms
16 fresh asparagus spears
¾ c. water
12 peeled uncooked shrimp
¼ c. unsalted butter
1 red pepper, cut into strips
1 tbsp. olive oil
2 tsp. lemon juice

Arrange spinach on 4 salad plates. Sprinkle mushrooms over spinach. Cut asparagus into ½-inch pieces. Combine with water in saucepan. Cook until tender-crisp. Drain, reserving ½ cup liquid. Cut shrimp into thirds. Stir-fry in butter in skillet just until cooked through. Add red pepper. Stir-fry until tender-crisp. Add asparagus. Cook just until heated through. Remove to salad plates with slotted spoon. Stir reserved asparagus liquid into pan juices. Add olive oil. Cook until reduced to ⅓ cup liquid. Remove from heat. Whisk in lemon juice and salt and freshly ground pepper to taste. Spoon over salads. Yield: 4 servings.

Approx per serving: Cal 182, Prot 7.6 gr, T Fat 15.4 gr, Chol 61.5 mg, Carbo 5.7 gr, Sod 187.3 mg, Potas 383.9 mg.

Nancy Dunn
San Diego, CA

Tuna Salad

1   7-oz. can water-pack tuna,
   drained
1 lg. carrot, grated
½ c. chopped celery
¼ c. mayonnaise
¼ c. country-style salad dressing
1   3-oz. can chow mein noodles

Combine tuna, carrot, celery, mayonnaise and salad dressing in bowl; mix well. Chill for 1 hour. Toss with noodles just before serving. Yield: 4 servings.
Note: Nutritional information does not include country-style salad dressing.

Approx per serving: Cal 273, Prot 16.2 gr, T Fat 16.6 gr, Chol 41.3 mg, Carbo 15.0 gr, Sod 726.0 mg, Potas 261.2 mg.

Mary Anne Fleischacker
Columbus H.S., Columbus, NE

Tuna Salad Vinaigrette

½ c. olive oil
3 tbsp. lemon juice
2 cloves of garlic, crushed
Freshly ground black pepper to taste
1 tsp. basil
1 bunch scallions, minced
1   8-oz. package spiral pasta, cooked
2   7-oz. cans water-pack tuna, drained
1   16-oz. can green beans, drained
4 lg. tomatoes, sliced

Combine olive oil, lemon juice, garlic, black pepper and basil in bowl; mix well. Combine scallions, pasta, tuna and green beans in bowl. Pour dressing over top; toss to coat. Place overlapping tomato slices on serving plate. Spoon tuna mixture onto tomatoes. Sprinkle with additional freshly ground black pepper. Yield: 4 servings.

Approx per serving: Cal 522, Prot 36.0 gr, T Fat 15.5 gr, Chol 58.0 mg, Carbo 59.6 gr, Sod 972.5 mg, Potas 895.0 mg.

Mary Hyder
Plattsburg, NY

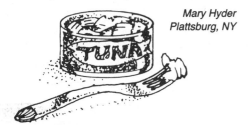

Ham and Asparagus Salad

1 lb. fresh or frozen asparagus
2 c. chopped ham
8 mushrooms, sliced
½ c. shredded Swiss cheese
1 sm. onion, chopped
2 tbsp. sesame seed
¾ c. Italian salad dressing
6 lettuce cups
1 tomato, cut into wedges

Cut asparagus into bite-sized pieces. Cook in water to cover in saucepan until tender-crisp; drain. Chill in refrigerator. Combine with ham, mushrooms, cheese, onion and sesame seed in bowl. Add salad dressing; mix lightly. Spoon into lettuce cups. Top with tomato wedges. Yield: 6 servings.

Approx per serving: Cal 385, Prot 15.7 gr, T Fat 32.9 gr, Chol 51.0 mg, Carbo 8.7 gr, Sod 1049.6 mg, Potas 419.5 mg.

Wilma Beatty
Chicago, IL

Taos Salad Toss

1 c. French salad dressing
1 tbsp. chopped green chilies
1 tsp. instant minced onion
¾ tsp. chili powder
¼ tsp. salt
1 med. head lettuce, chopped
1   15-oz. can kidney beans,
   drained
½ c. sliced ripe olives
1 med. tomato, cut into wedges
½ c. shredded sharp Cheddar cheese
½ c. tortilla chips

Combine first 5 ingredients in bowl; mix well. Chill until serving time. Mix lettuce, beans and olives in salad bowl. Chill until serving time. Spoon dressing into center of lettuce mixture, arrange tomato wedges in circle, sprinkle cheese over top and trim edge of bowl with tortilla chips. Toss salad just before serving. Yield: 8 servings.

Approx per serving: Cal 296, Prot 7.1 gr, T Fat 20.3 gr, Chol 9.8 mg, Carbo 23.9 gr, Sod 700.5 mg, Potas 369.6 mg.

Chuck Frey
Hemphill Co., TX

Parsley Salad

½ c. Lebanese cracked wheat
1 lg. bunch parsley, minced
2 to 3 tomatoes, peeled, finely chopped
1 tbsp. minced onion
¼ c. olive oil
Juice of 1 lemon

Soak wheat in hot water to cover in bowl for 1 hour; drain. Combine with remaining ingredients; mix well. Chill for 1 hour. Yield: 4 servings.

Approx per serving: Cal 227, Prot 4.0 gr,
T Fat 14.2 gr, Chol 0.0 mg, Carbo 23.1 gr,
Sod 10.8 mg, Potas 408.0 mg.

Ardis G. Landauer
San Diego, CA

Rice and Lentil Salad

2 c. rice, cooked
1 c. lentils, cooked
½ c. chopped tomato
½ c. chopped green pepper
½ c. chopped chives
½ c. chopped celery
½ c. chopped carrots
½ c. chopped broccoli
½ c. chopped cauliflower
½ c. chopped mushrooms
½ c. chopped radishes
½ c. chopped avocado
1 8-oz. bottle of Italian dressing

Mix cooled rice and lentils in bowl. Add remaining vegetables. Pour dressing over top. Chill for 1 to 2 hours. Yield: 8 servings.

Approx per serving: Cal 455, Prot 10.5 gr,
T Fat 20.9 gr, Chol 0.0 mg, Carbo 58.0 gr,
Sod 657.3 mg, Potas 532.5 mg.

Barbara Lewis
Little Rock, AR

Sunny-Day Salad

1 c. cooked rice
2 c. shredded cabbage
¼ c. shredded pineapple
¼ c. shredded carrots
¼ c. chopped walnuts
¼ c. raisins
¼ c. mayonnaise

Combine first 6 ingredients in bowl. Stir in mayonnaise. Chill for 1 hour. Serve on lettuce leaves. Yield: 3 servings.

Approx per serving: Cal 315, Prot 3.8 gr,
T Fat 21.5 gr, Chol 13.1 mg, Carbo 30.1 gr,
Sod 249.7 mg, Potas 335.0 mg.

Shaundra Haislett
Indianapolis, IN

Vermicelli Salad

2 pkg. dry Italian salad dressing mix
2 c. low-fat milk
2 c. reduced-calorie mayonnaise
16 oz. vermicelli, broken
2 c. chopped celery
½ c. chopped onion
3 hard-boiled eggs, chopped

Combine dressing mix, milk and mayonnaise in bowl; mix well. Cook vermicelli according to package directions; drain. Add to dressing; mix well. Fold in remaining ingredients. Chill, covered, for 48 to 76 hours. Yield: 10 servings. Note: Nutritional information does not include salad dressing mix.

Approx per serving: Cal 364, Prot 9.6 gr,
T Fat 18.8 gr, Chol 93.8 mg, Carbo 41.2 gr,
Sod 90.8 mg, Potas 280.2 mg.

Maxine Stutts
Pontiac, MI

Asparagus Garden Salad

½ c. cooked asparagus tips
½ c. sliced radishes
½ c. cucumber
1 c. shredded lettuce
2 tbsp. minced green pepper
4 green onions, minced
1 tbsp. minced parsley
¼ c. grated Cheddar cheese
¼ c. French dressing

Combine all ingredients except dressing in bowl. Add dressing; toss to coat. Yield: 4 servings.

Approx per serving: Cal 111, Prot 2.8 gr,
T Fat 8.6 gr, Chol 7.0 mg, Carbo 6.7 gr,
Sod 316.7 mg, Potas 173.8 mg.

Donna Flewallen
Weatherford, TX

Four-Bean Salad

6 tbsp. sugar
3 tbsp. oil
⅓ c. tarragon vinegar
½ tsp. salt
¼ tsp. pepper
1　15-oz. can kidney beans, drained
1　15-oz. can wax beans, drained
1　15-oz. can French-cut green beans,
*　drained*
1 can garbanzo beans, drained
1 sm. onion, chopped
1 sm. green pepper, chopped

Combine first 5 ingredients in bowl; mix well. Add beans, onion and green pepper; mix well. Chill in refrigerator for 12 hours or longer.
Yield: 12 servings.

Approx per serving: Cal 167, Prot 6.6 gr,
T Fat 4.5 gr, Chol 0.0 mg, Carbo 26.8 gr,
Sod 149.0 mg, Potas 314.0 mg.

Barbara Brasher
Wetzel, WV

Fresh Green Bean-Mushroom Salad

1 lb. fresh green beans
1½ c. sliced mushrooms
½ c. reduced-calorie Italian dressing
½ c. onion rings
¼ c. chopped ripe olives

Cut green beans into 2-inch pieces. Cook in a small amount of water in saucepan until tender; drain. Combine with mushrooms and Italian dressing in bowl; mix well. Chill for several hours. Spoon into lettuce-lined serving bowl. Top with onion rings and ripe olives. Yield: 6 servings.

Approx per serving: Cal 55, Prot 2.2 gr,
T Fat 2.3 gr, Chol 0.0 mg, Carbo 8.1 gr,
Sod 208.9 mg, Potas 282.9 mg.

Sybil B. Murphy
Northwood H.S., Pittsboro, NC

Sweet and Sour Green Beans

1 lb. fresh green beans
1 onion, sliced into rings
¾ c. vinegar
¼ c. oil

¾ c. sugar
1 tsp. celery seed
½ tsp. salt
¼ tsp. pepper

Cut beans into 1-inch pieces. Combine with water to cover in medium saucepan. Cook until tender-crisp; drain. Combine with onion rings in bowl. Mix remaining ingredients in small bowl. Pour over beans; mix well. Chill, covered, for several hours to overnight. Yield: 8 servings.

Approx per serving: Cal 162, Prot 1.4 gr,
T Fat 6.9 gr, Chol 0.0 mg, Carbo 25.9 gr,
Sod 139.7 mg, Potas 194.2 mg.

Diane Zook
Larned H.S., Larned, KS

Beet and Celery Salad

2 c. chopped cooked beets
½ c. chopped unpeeled cucumber
2 tbsp. chopped sweet pickle
½ c. chopped celery
2 tbsp. chopped onion
1 tsp. sugar
⅓ c. reduced-calorie French dressing

Combine first 5 ingredients and salt and pepper to taste in bowl. Add mixture of sugar and French dressing. Chill for 30 minutes. Spoon onto lettuce-lined salad plates. Garnish with additional cucumber. Yield: 4 servings.

Approx per serving: Cal 69, Prot 1.3 gr,
T Fat 1.1 gr, Chol 0.0 mg, Carbo 15.1 gr,
Sod 415.0 mg, Potas 254.9 mg.

Allison Morris
Hershey, PA

Broccoli-Cauliflower Salad

4 c. broccoli flowerets
1 c. cauliflowerets
8 slices crisp-fried bacon, crumbled
2 sm. onions, chopped
½ c. raisins
¾ c. mayonnaise
¼ c. sugar
3 tbsp. cider vinegar

Combine broccoli, cauliflower, bacon, onions and raisins in bowl; mix well. Mix remaining ingredients in small bowl. Pour over vegetables; toss to mix. Chill. Yield: 6 servings.

Approx per serving: Cal 367, Prot 6.8 gr, T Fat 27.9 gr, Chol 28.5 mg, Carbo 26.7 gr, Sod 288.7 mg, Potas 508.1 mg.

Sarah E. Etheredge
Carroll H.S., Ozark, AL

Broccoli Mixed Salad

2 c. each chopped cauliflower, broccoli
½ c. chopped onion
4 carrots, chopped
2 tbsp. salad dressing
¼ c. buttermilk
1 tsp. ranch-style salad dressing mix

Combine vegetables in bowl. Mix salad dressing, buttermilk and dressing mix in small bowl. Pour over vegetables. Chill for several hours.
Yield: 6 servings.

Approx per serving: Cal 82, Prot 3.6 gr, T Fat 2.5 gr, Chol 2.7 mg, Carbo 13.3 gr, Sod 88.9 mg, Potas 515.0 mg.

Beverly Nickel
Newton H.S., Newton, KS

Marinated Broccoli Salad

Flowerets of 1 bunch broccoli
1   7-oz. can pitted ripe olives, sliced
1   2-oz. jar mushrooms, drained
1 c. chopped celery
1   8-oz. can sliced water chestnuts, drained
3 green onions, sliced
1   8-oz. bottle of Italian dressing
2 med. tomatoes, cut into wedges

Combine broccoli, olives, mushrooms, celery, water chestnuts and green onions in bowl. Add Italian dressing; mix well. Chill, tightly covered, overnight. Top with tomato wedges at serving time. Yield: 8 servings.

Approx per serving: Cal 229, Prot 3.0 gr, T Fat 21.7 gr, Chol 0.0 mg, Carbo 8.9 gr, Sod 782.8 mg, Potas 365.1 mg.

Linda K. Turner
Northeastern Jr.-Sr. H.S., Richmond, IN

Cabbage Patch Coleslaw

3 c. shredded cabbage
½ c. chopped parsley
½ c. sliced green onions
3 tbsp. sugar
3 tbsp. vinegar
2 tbsp. oil
1 tsp. salt

Mix cabbage, parsley and green onions in salad bowl. Combine remaining ingredients in small bowl; mix well. Pour over vegetables. Toss lightly to coat well. Garnish with sprigs of parsley.
Yield: 6 servings.

Approx per serving: Cal 81, Prot 0.9 gr, T Fat 4.7 gr, Chol 0.0 mg, Carbo 10.3 gr, Sod 367.1 mg, Potas 167.4 mg.

Tamara Roberts
Anniston, GA

California Salad

Flowerets of 1 med. head cauliflower
Flowerets of 2 stalks broccoli
1 bunch green onions, chopped
¼ c. green olives
¼ c. chopped green pepper
¼ c. chopped sweet red pepper
½ c. cubed longhorn cheese
1 c. salad dressing
2 tbsp. vinegar
2 tbsp. sugar
½ tsp. salt

Combine cauliflower, broccoli, green onions, olives, green and red peppers and cheese in bowl; mix well. Mix remaining ingredients in bowl. Pour over vegetables. Chill for several hours. Yield: 8 servings.

Approx per serving: Cal 166, Prot 7.2 gr, T Fat 9.7 gr, Chol 14.5 mg, Carbo 16.4 gr, Sod 410.8 mg, Potas 593.3 mg.

Evelyn Marvin
Fowler H.S., Fowler, CO

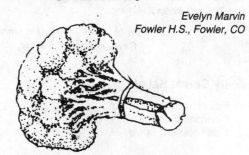

Carrot Copper Pennies

2 lb. fresh carrots, sliced ¼ in. thick
2 med. onions, thinly sliced into rings
1 med. green pepper, cut into thin strips
1 can tomato soup
¾ c. vinegar
½ c. oil
1 tsp. Worcestershire sauce
⅔ c. sugar
1 tsp. mustard
½ tsp. salt

Cook carrots in a small amount of water in saucepan for 8 to 10 minutes or until tender-crisp; drain. Combine with onion and green pepper in bowl. Mix remaining ingredients in small bowl. Pour over vegetables; mix well. Chill overnight. Drain, reserving marinade. Store unused salad in reserved marinade for up to several days. Yield: 10 servings.

Approx per serving: Cal 222, Prot 2.0 gr,
T Fat 11.8 gr, Chol 0.0 mg, Carbo 29.8 gr,
Sod 405.8 mg, Potas 445.5 mg.

Betty K. Munsey
Bland H.S., Bland, VA

Creamy Carrot-Nut Mold

1 6-oz. package orange gelatin
2 c. boiling water
1 8-oz. carton sour cream
2 c. grated carrots
1½ c. crushed pineapple
½ c. chopped pecans

Dissolve gelatin in hot water in bowl. Add sour cream; stir until smooth. Add remaining ingredients; mix well. Pour into mold. Chill until set. Unmold onto serving plate.
Yield: 10 servings.

Approx per serving: Cal 190, Prot 3.2 gr,
T Fat 9.1 gr, Chol 10.1 mg, Carbo 26.4 gr,
Sod 77.0 mg, Potas 216.1 mg.

Kirby McHenry
Pascagoula H.S., Gautier, MS

Zesty Carrot Salad

5 c. sliced carrots
1 onion, sliced
1 can tomato soup

¾ c. vinegar
½ c. oil
1 c. sugar
1 tsp. Worcestershire sauce
1 tsp. mustard
1 tsp. salt
½ tsp. pepper

Cook carrots in a small amount of water in saucepan until tender; drain. Combine with remaining ingredients in bowl; mix well. Chill in refrigerator. Yield: 10 servings.

Approx per serving: Cal 228, Prot 1.4 gr,
T Fat 11.7 gr, Chol 0.0 mg, Carbo 31.8 gr,
Sod 493.8 mg, Potas 294.9 mg.

Carolyn Thompson
Redwater H.S., Texarkana, TX

Italian Cauliflower Salad

2 c. sliced cauliflower
½ c. sliced ripe olives
⅓ c. finely chopped green pepper
¼ c. chopped pimento
3 tbsp. chopped onion
4½ tbsp. olive oil
1½ tbsp. lemon juice
7½ tbsp. wine vinegar
¼ c. sugar
1 tsp. salt
Pinch of pepper

Combine cauliflower, olives, green pepper, pimento and onion in bowl. Mix remaining ingredients in small bowl. Pour over vegetables; mix well. Chill for 1 hour or longer.
Yield: 4 servings.

Approx per serving: Cal 240, Prot 1.8 gr,
T Fat 18.8 gr, Chol 0.0 mg, Carbo 19.4 gr,
Sod 671.4 mg, Potas 237.3 mg.

Ann McMullin
Smith-Cotton H.S., Sedalia, MO

Caesar Potato Salad

4 potatoes, peeled, cooked, chopped
4 pitted ripe olives, sliced
¼ c. reduced-calorie Italian
**   salad dressing**
1 egg, beaten

1 tbsp. Worcestershire sauce
2 tsp. mustard
8 tsp. Parmesan cheese

Combine potatoes and olives in bowl. Mix remaining ingredients and garlic salt to taste in small bowl until smooth. Add to potatoes; mix well. Chill, covered, in refrigerator. Yield: 8 servings.

Approx per serving: Cal 101, Prot 3.8 gr, T Fat 2.2 gr, Chol 33.9 mg, Carbo 16.8 gr, Sod 135.7 mg, Potas 411.5 mg.

Lisa Davis
Boston, MA

Herbed Potato Salad

½ c. low-fat yogurt
¼ c. reduced-calorie mayonnaise
2 tbsp. chopped parsley
1 tbsp. chopped fresh basil
1 tbsp. chopped green onion
3 med. potatoes, cooked, chopped
½ c. frozen green peas, thawed

Combine yogurt, mayonnaise, parsley, basil and green onion in bowl; mix well. Add potatoes and peas; mix lightly. Chill, covered, for several hours. Yield: 4 servings.

Approx per serving: Cal 329, Prot 5.6 gr, T Fat 6.2 gr, Chol 7.3 mg, Carbo 71.0 gr, Sod 58.6 mg, Potas 999.0 mg.

Provi Presley
Morton, MS

Fresh Spinach Salad

1 lb. fresh spinach, torn
1 med. red onion, sliced into rings
4 stalks celery, chopped
8 mushrooms, sliced
4 green onions, chopped
¼ c. raisins
4 tsp. chopped parsley
¼ c. chopped walnuts
4 slices crisp-fried bacon, crumbled
¼ c. reduced-calorie Italian
*   salad dressing*
½ c. orange juice
4 hard-boiled eggs, cut into quarters
½ c. garlic-flavored croutons

Combine first 9 ingredients in bowl; mix well. Mix salad dressing and orange juice in small bowl. Pour over salad; toss lightly. Top with egg quarters and croutons. Yield: 8 servings.

Approx per serving: Cal 147, Prot 8.1 gr, T Fat 8.2 gr, Chol 132.1 mg, Carbo 12.3 gr, Sod 220.6 mg, Potas 525.9 mg.

Ann Gillis
Athens, GA

Fire and Ice Tomatoes

6 lg. tomatoes, peeled, quartered
2 lg. green peppers, cut into strips
1 lg. red onion, sliced
¾ c. vinegar
1½ tsp. each celery salt, mustard seed
½ tsp. salt
⅛ tsp. each pepper, cayenne pepper
4½ tbsp. sugar

Layer tomatoes, green peppers and onion in bowl. Combine remaining ingredients in saucepan. Bring to a boil. Pour over vegetables. Chill, covered, for 24 hours. Yield: 12 servings.

Approx per serving: Cal 42, Prot 1.0 gr, T Fat 0.2 gr, Chol 0.0 mg, Carbo 10.2 gr, Sod 360.1 mg, Potas 213.3 mg.

Kathy Gerace
Detroit, MI

Confetti Salad

½ c. grated peeled carrot
1 tomato, peeled, chopped
⅓ c. each chopped cucumber, celery
2 tbsp. chopped green pepper
2 tbsp. chopped green onion
2 c. cottage cheese
1 tsp. salt

Combine vegetables in bowl; mix well. Add cottage cheese and salt; toss gently to mix. Serve in lettuce cups or lettuce-lined salad bowl. Yield: 6 servings.

Approx per serving: Cal 99, Prot 11.6 gr, T Fat 3.5 gr, Chol 15.5 mg, Carbo 5.0 gr, Sod 556.5 mg, Potas 197.2 mg.

Sarah Peterson
Bernalillo County, NM

Marinated Vegetables

Flowerets of 1 med. head cauliflower
Flowerets of 1 med. bunch broccoli
3 med. zucchini, diagonally sliced
1 pt. cherry tomatoes
2 c. whole fresh mushrooms
1 7-oz. can pitted black olives, drained
2 tsp. salt
½ tsp. pepper
2 cups reduced-calorie Italian dressing

Combine all ingredients in large bowl; mix well. Chill, covered, for 6 hours to overnight. Yield: 12 servings.

Approx per serving: Cal 86, Prot 4.2 gr, T Fat 4.5 gr, Chol 0.0 mg, Carbo 10.3 gr, Sod 771.0 mg, Potas 510.2 mg.

Emma Ellen Bunyard
Jenks H.S., Jenks, OK

Twenty-Four Hour Vegetable Salad

¼ lb. spinach, torn
½ lb. lettuce, torn
¼ c. chopped green pepper
¾ c. chopped celery
2 tbsp. chopped red onion
½ 10-oz. package frozen peas
1 tsp. sugar
⅔ c. salad dressing
¼ c. each grated cheese, bacon bits

Combine spinach and lettuce in salad bowl. Layer green pepper, celery, onion and peas over greens. Sprinkle with sugar. Spread salad dressing over layers, sealing to edge. Top with cheese and bacon bits. Chill overnight. Toss to serve if desired. Yield: 10 servings.

Approx per serving: Cal 117, Prot 3.6 gr, T Fat 8.4 gr, Chol 10.8 mg, Carbo 7.5 gr, Sod 342.8 mg, Potas 213.7 mg.

Debra S. Hart
Emerson Jr. H.S., Enid, OK

Low-Fat Yogurt Dressing

1 c. low-fat yogurt
¼ c. lemon juice
1 tbsp. chopped chives
1 tsp. minced parsley
½ tsp. paprika

¼ tsp. dry mustard
Pinch of cayenne pepper

Combine all ingredients in small bowl; mix well. Add salt to taste. Chill until serving time. Serve on lettuce and tomato salad. Yield: 1½ cups.

Approx per tablespoon: Cal 5, Prot 0.3 gr, T Fat 0.2 gr, Chol 0.8 mg, Carbo 0.7 gr, Sod 4.8 mg, Potas 17.0 mg.

Ann McMullin
Smith Cotton H.S., Sedalia, MO

Special Dressing for Green Salads

1 c. oil
1 can tomato soup
⅓ c. wine vinegar
1 c. mayonnaise
½ c. packed brown sugar
1 tsp. each dry mustard, garlic salt
½ tsp. each onion salt, celery salt
¼ tsp. Worcestershire sauce
¼ tsp. paprika
Dash of Tabasco sauce

Process all ingredients in blender container at low speed for several seconds. Process at high speed until smooth. Chill, covered, until serving time. Yield: 50 tablespoons.

Approx per tablespoon: Cal 84, Prot 0.1 gr, T Fat 8.1 gr, Chol 3.1 mg, Carbo 3.1 gr, Sod 161.3 mg, Potas 22.4 mg.

Kay Caskey
Swope Mid. Sch., Reno, NV

Dill Dip for Fresh Vegetables

1 c. sour cream
½ c. mayonnaise
1 tbsp. chopped green onion
2 tsp. parsley flakes
1 tsp. each dill, seasoned salt

Combine all ingredients in serving dish. Chill, covered, until serving time. Yield: 1½ cups.

Approx per tablespoon: Cal 25, Prot 0.4 gr, T Fat 2.3 gr, Chol 4.4 mg, Carbo 0.9 gr, Sod 95.8 mg, Potas 23.5 mg.

Joyce R. Daglow
Waldron Area Sch., Waldron, MI

MAIN DISHES

Skillet Beef and Broccoli

2 c. thin round steak strips
1 tbsp. soy sauce
1 tbsp. oil
1 to 2 tbsp. sesame seed
4 c. broccoli flowerets
½ c. beef broth
Garlic powder to taste
1 c. sliced mushrooms
1 tbsp. cornstarch
1 tbsp. lemon juice
½ c. beef broth
1   6-oz. can water chestnuts, drained

Combine steak, soy sauce and oil in bowl. Let stand for several minutes. Heat electric skillet to 300 degrees. Add sesame seed. Cook for 1 to 1½ minutes or until toasted. Remove and reserve sesame seed. Add broccoli and ½ cup broth to skillet. Cook until tender. Increase temperature to 375 degrees. Add steak and garlic powder. Cook until brown, stirring frequently. Add mushrooms. Cook until tender. Blend cornstarch with lemon juice and remaining ½ cup broth. Stir into skillet with water chestnuts. Cook until thickened, stirring constantly. Spoon into serving dish. Sprinkle with reserved sesame seed. Serve over bean sprouts, rice or chow mein noodles. Yield: 4 servings.

Approx per serving: Cal 373, Prot 31.1 gr,
T Fat 20.7 gr, Chol 77.9 mg, Carbo 18.1 gr,
Sod 857.5 mg, Potas 952.6 mg.

Susan Shay
Hidalgo County, NM

Beef and Broccoli Stir-Fry

½ lb. beef round steak
1 tbsp. cornstarch
1 tbsp. soy sauce
1 tsp. sugar
2 tsp. minced gingerroot
1 clove of garlic, minced
1 tbsp. peanut oil
1 lb. broccoli, chopped
1 onion, chopped
2 carrots, diagonally sliced
2 tbsp. peanut oil
¾ oz. cashews
1 tbsp. cornstarch
3 tbsp. soy sauce
½ c. water

Slice beef cross grain into thin slices. Marinate in mixture of 1 tablespoon cornstarch, 1 tablespoon soy sauce, sugar, gingerroot and garlic for 15 minutes. Stir-fry beef in 1 tablespoon hot oil in wok for 1 minute; remove beef. Stir-fry broccoli, onion and carrots in 2 tablespoons oil for 4 minutes or until tender-crisp. Add beef, cashews and mixture of 1 tablespoon cornstarch, 3 tablespoons soy sauce and water. Cook until thickened, stirring constantly. Serve over rice. Yield: 4 servings.

Approx per serving: Cal 314, Prot 17.5 gr,
T Fat 19.2 gr, Chol 36.9 mg, Carbo 20.8 gr,
Sod 676.0 mg, Potas 793.4 mg.

Patsy Tucker
Austin, TX

Quick Beef and Cauliflower

1 lb. round steak, cut into ½-in. cubes
2 tbsp. oil
1 med. head cauliflower, broken into
*   flowerets*
1 green pepper, cut into ¾-in. pieces
1 clove of garlic, minced
¼ c. soy sauce
2 tbsp. cornstarch
½ tsp. sugar
1½ c. beef broth
1 c. sliced green onions
8 oz. mushrooms, sliced
3 c. hot cooked rice

Brown steak cubes in oil in skillet. Add cauliflower, green pepper, garlic and soy sauce; mix well. Simmer, covered, for 8 to 10 minutes. Blend cornstarch, sugar and broth in bowl. Stir into beef mixture with green onions and mushrooms. Cook until thickened, stirring constantly. Serve over rice. Yield: 6 servings.

Approx per serving: Cal 372, Prot 24.3 gr,
T Fat 13.3 gr, Chol 55.2 mg, Carbo 40.0 gr,
Sod 1521.5 mg, Potas 918.3 mg.

Sandra Lane
Baton Rouge, LA

Beef and Vegetables Sauté

1   4-oz. can mushrooms
8 oz. lean beef, cut into strips
2 tbsp. oil

1 med. onion, chopped
1 c. cut fresh green beans
1 c. each sliced green pepper, celery
4 tsp. cornstarch
1 tbsp. soy sauce

Drain mushrooms reserving liquid. Add enough water to liquid to measure ¾ cup. Sauté beef strips in hot oil in skillet. Add vegetables. Sauté for 3 to 5 minutes or until tender-crisp. Add mixture of reserved liquid, cornstarch, soy sauce, mushrooms and salt and pepper to taste. Cook until thickened, stirring constantly. Yield: 4 servings.

Approx per serving: Cal 219, Prot 13.5 gr,
T Fat 14.0 gr, Chol 38.6 mg, Carbo 10.1 gr,
Sod 415.0 mg, Potas 458.2 mg.

Frankie Stuart
Little Rock, AR

Beef and Swiss Spread

2 c. ground cooked roast beef
1 c. shredded Swiss cheese
½ c. sweet pickle relish
2 tbsp. finely chopped onion
¾ c. mayonnaise
¼ tsp. salt

Combine all ingredients in bowl; mix well. Chill in refrigerator. Spread on bread or toasted buns or spoon into pita rounds. Yield: 8 servings.

Approx per serving: Cal 293, Prot 12.2 gr,
T Fat 24.7 gr, Chol 53.9 mg, Carbo 6.0 gr,
Sod 413.6 mg, Potas 119.5 mg.

Linda Brown
Trenton, NJ

Rib Eye Stir-Fry

1½ lb. rib eye steaks, cut into thin strips
½ c. sukiyaki sauce
2 cloves of garlic, thinly sliced
¼ c. peanut oil
2 c. each thinly sliced celery, mushrooms
1 bunch green onions, sliced
3 tbsp. oyster sauce

Marinate steak strips in sukiyaki sauce for 1 hour, stirring occasionally; drain. Stir-fry steak and garlic in hot oil in wok until brown. Add vegetables. Stir-fry for 2 to 3 minutes or until tender-

crisp. Add oyster sauce. Heat to serving temperature. Serve over rice. Yield: 4 servings.
Note: Nutritional information does not include sukiyaki sauce or oyster sauce.

Approx per serving: Cal 364, Prot 24.8 gr,
T Fat 24.6 gr, Chol 72.3 mg, Carbo 11.4 gr,
Sod 164.3 mg, Potas 815.3 mg.

Linda Miles
Clinton, UT

Broiled Marinated Flank Steak

¼ c. oil
2 tbsp. each catsup, soy sauce, water
1 sm. onion, finely chopped
1 clove of garlic, minced
1 1-lb. flank steak, scored

Mix first 6 ingredients in plastic bag. Add steak; tie bag. Marinate overnight in refrigerator, turning frequently. Drain, reserving marinade. Broil to desired degree of doneness, basting with marinade. Slice cross grain. Yield: 4 servings.

Approx per serving: Cal 307, Prot 25.5 gr,
T Fat 20.2 gr, Chol 77.1 mg, Carbo 4.8 gr,
Sod 825.4 mg, Potas 189.4 mg.

Miriam Jacobson
Buffalo, NY

Microwave Stir-Fry

½ lb. round steak, thinly sliced
1 onion, sliced
1 pkg. Chinese pea pods
5 or 6 fresh mushrooms
½ c. bean sprouts
1 tbsp. (about) soy sauce

Place steak and onion slices in preheated microwave browning dish. Microwave on High for 3 minutes. Add remaining ingredients. Microwave for 4 minutes; stir. Let stand for 3 minutes before serving. Yield: 2 servings.
Note: Nutritional information does not include pea pods.

Approx per serving: Cal 298, Prot 43.6 gr,
T Fat 8.1 gr, Chol 116.0 mg, Carbo 11.5 gr,
Sod 772.1 mg, Potas 816.8 mg.

Helen D. Tunison
Adams, PA

Beef Burgundy

½ c. flour
1 tsp. salt
½ tsp. pepper
¼ tsp. cayenne pepper
2 lb. beef sirloin, cut into cubes
¼ c. margarine
2 tbsp. Brandy
18 to 24 sm. white onions, peeled
½ lb. medium mushrooms
1 clove of garlic, crushed
6 lg. shallots, finely chopped
2 tbsp. chopped parsley
¼ tsp. thyme
1 bay leaf
1 c. condensed beef broth
3 tbsp. beef broth concentrate
1 c. Burgundy

Combine first 5 ingredients in plastic bag. Shake to coat steak; reserve remaining flour. Brown beef ¼ at a time in hot margarine in large skillet. Place beef in 2½-quart casserole. Heat Brandy in small saucepan. Ignite and pour over beef. Add onions, mushrooms and reserved flour; mix well. Stir garlic, shallots, parsley and thyme into pan drippings. Cook for 1 minute. Add next 3 ingredients. Bring to a boil, stirring to deglaze pan. Pour over beef and vegetables. Bake, covered, at 350 degrees for 1½ hours. Let stand for 20 minutes. Serve over wild rice prepared according to package directions, substituting Burgundy for half the water. Yield: 8 servings.
Note: Nutritional information does not include wild rice.

Approx per serving: Cal 572, Prot 25.6 gr,
T Fat 36.4 gr, Chol 84.1 mg, Carbo 30.3 gr,
Sod 1150.6 mg, Potas 859.2 mg.

Stacey Leigh Hillis
Midland, TX

Fajitas

1 lb. top round steak, cut into thin strips
3 tbsp. red wine vinegar
3 tbsp. oil
1 tsp. garlic salt
½ tsp. cumin
12   6-in. flour tortillas
Juice of ½ lime

Combine steak strips with vinegar, oil, garlic salt and cumin in bowl; mix well. Marinate for 20 minutes or longer. Wrap tortillas in damp cloth. Warm in 350-degree oven for 10 to 15 minutes. Drain steak strips. Stir-fry in skillet for 1 to 2 minutes or until tender. Sprinkle with lime juice. Spoon onto warm tortillas. Top with Guacamole and Picante Sauce. Roll to enclose filling.
Yield: 12 servings.

Approx per serving: Cal 201, Prot 10.1 gr,
T Fat 9.9 gr, Chol 25.7 mg, Carbo 17.8 gr,
Sod 204.4 mg, Potas 128.6 mg.

Guacamole

2 med. avocados, mashed
1 tbsp. mayonnaise
Juice of ½ lemon
¾ tsp. seasoned salt

Combine all ingredients in bowl; mix until smooth. Yield: 32 tablespoons.

Approx per tablespoon: Cal 27, Prot 0.3 gr,
T Fat 2.7 gr, Chol 0.3 mg, Carbo 1.0 gr,
Sod 53.1 mg, Potas 87.0 mg.

Picante Sauce

1   16-oz. can stewed tomatoes
1 sm. onion, chopped
1 tbsp. chopped jalapeño pepper
1 tbsp. garlic powder
1 tsp. salt
½ tsp. sugar
6 cilantro leaves
1 tbsp. cider vinegar

Process all ingredients in blender container for 30 seconds or until smooth. Store in covered container in refrigerator. Yield: 50 tablespoons.

Approx per serving: Cal 3, Prot 0.1 gr,
T Fat 0.0 gr, Chol 0.0 mg, Carbo 0.6 gr,
Sod 54.6 mg, Potas 23.0 mg.

Susan Crumley
Kingsport, TN

Steak Fingers

2 tbsp. margarine
1 egg, beaten
1 tsp. water
⅓ c. dry bread crumbs
2 tbsp. Parmesan cheese

Onion powder to taste
1 lb. round steak, cut into strips
2 tbsp. flour

Melt margarine in 9x13-inch baking dish. Beat egg with water in bowl. Mix bread crumbs and Parmesan cheese with onion powder to taste in bowl. Dip steak strips in flour, egg and crumb mixture. Arrange in prepared baking dish. Bake for 30 minutes, turning strips once. Place on serving plate. Yield: 4 servings.

Approx per serving: Cal 336, Prot 26.8 gr,
T Fat 20.6 gr, Chol 140.8 mg, Carbo 9.2 gr,
Sod 227.3 mg, Potas 290.7 mg.

Melodye Farley
Frankfort, KY

Green Pepper Steak

1 lb. round steak, partially frozen
¼ c. reduced-sodium soy sauce
1 tbsp. cornstarch
½ tsp. sugar
¼ tsp. ginger
2 tbsp. oil
3 med. green peppers,
*   cut into 1-in. pieces*
1 clove of garlic, minced
¼ c. water
2 sm. tomatoes, cut into wedges

Slice steak diagonally cross grain into thin strips. Place in shallow bowl. Blend soy sauce into mixture of cornstarch, sugar and ginger in bowl. Pour over steak; mix well. Stir-fry steak ⅓ at a time in hot oil in skillet until brown. Remove with slotted spoon when brown. Reduce heat. Add green peppers, garlic and water. Cook for 5 minutes or until peppers are tender-crisp. Add steak and tomatoes. Cook until heated through, stirring constantly. Spoon onto serving plate. Yield: 4 servings.
Note: Nutritional information does not include reduced-sodium soy sauce.

Approx per serving: Cal 314, Prot 24.5 gr,
T Fat 19.2 gr, Chol 73.8 mg, Carbo 10.8 gr,
Sod 71.6 mg, Potas 644.9 mg.

Mabel Maxwell
Detroit, MI

Skinny Beef Stroganoff

1 lb. round steak
1 tbsp. oil
8 oz. mushrooms, sliced
1 env. dry onion soup mix
1 tsp. dillweed
⅔ c. water
1 c. yogurt
2 tbsp. flour

Slice steak diagonally cross grain into ⅛-inch strips. Stir-fry steak in hot oil in skillet. Remove steak and pan juices. Stir-fry mushrooms for 2 to 3 minutes. Add soup mix, dillweed and water. Simmer for 3 minutes. Add steak with juices. Stir in mixture of yogurt and flour. Cook until thickened, stirring constantly. Serve over noodles or rice. Yield: 4 servings.

Approx per serving: Cal 341, Prot 27.7 gr,
T Fat 18.9 gr, Chol 94.6 mg, Carbo 14.3 gr,
Sod 810.1 mg, Potas 630.9 mg.

Dabney Adams
Philadelphia, PA

Mushroom and Pepper Steak

1 lb. round steak
2 tbsp. paprika
1 onion, chopped
2 tbsp. oil
2 to 3 tbsp. garlic powder
2 beef bouillon cubes
2 to 3 c. water
1   4-oz. can mushrooms, drained
1 med. green pepper, chopped
1 tbsp. cornstarch
2 tbsp. soy sauce

Slice steak diagonally cross grain into thin strips. Sprinkle with paprika. Brown steak and onion in oil in skillet. Add garlic powder, bouillon cubes and water. Simmer for 30 to 45 minutes or until of desired consistency. Add mushrooms and green pepper. Simmer for 10 minutes. Add mixture of cornstarch and soy sauce. Cook until thickened, stirring constantly. Serve over rice. Yield: 6 servings.

Approx per serving: Cal 204, Prot 16.3 gr,
T Fat 12.8 gr, Chol 50.2 mg, Carbo 5.3 gr,
Sod 800.6 mg, Potas 261.6 mg.

May Roark
Mountain City, TN

Oriental Pepper Steak

2 lb. beef chuck
3 tbsp. oil
2 beef bouillon cubes
½ tsp. sugar
1½ c. hot water
1 clove of garlic, minced
¼ tsp. pepper
2 green peppers, cut into strips
2 sweet red peppers, cut into strips
2 tbsp. cornstarch
2 tbsp. soy sauce
½ c. sliced green onions

Cut beef into thin strips. Brown quickly in hot oil in skillet. Place in Crock • Pot. Dissolve bouillon cubes and sugar in hot water. Pour over beef. Sprinkle with garlic and pepper. Cook on Low for 6 to 8 hours or until tender. Set temperature to High. Add green pepper strips and cornstarch blended with soy sauce. Cook for 35 minutes or until thickened. Serve with rice or noodles. Sprinkle with green onions. Yield: 8 servings.

Approx per serving: Cal 291, Prot 21.6 gr,
T Fat 19.9 gr, Chol 72.2 mg, Carbo 5.5 gr,
Sod 610.8 mg, Potas 276.7 mg.

Sarah E. Etheredge
Carroll H.S., Ozark, AL

Red Peppers and Steak

12 oz. boneless chuck steak
1 med. sweet red pepper, cut into strips
1 tbsp. oil
¾ c. canned beef broth
1 tsp. soy sauce
2 tbsp. cornstarch
1 tbsp. water
⅛ tsp. pepper

Broil steak on broiler pan until rare. Cut into thin strips. Sauté red pepper with oil in skillet. Add beef broth and soy sauce. Cook for 3 minutes. Blend cornstarch with 1 tablespoon water. Stir into steak mixture. Cook until mixture thickens, stirring constantly. Season with pepper. Spoon into serving bowl. Yield: 2 servings.

Approx per serving: Cal 328, Prot 28.1 gr,
T Fat 17.9 gr, Chol 92.9 mg, Carbo 12.3 gr,
Sod 971.0 mg, Potas 391.2 mg.

Carol Bender
Dallas, TX

Steak and Mushroom Bits

1½ lb. beef steak, sliced
¼ c. oil
1 med. onion, sliced
1 green pepper, coarsely chopped
1 clove of garlic, minced
1 lb. mushrooms, sliced
2 tbsp. flour
1 beef bouillon cube
¾ c. boiling water
1 8-oz. can tomatoes, chopped
2 tsp. Worcestershire sauce
1 tsp. salt

Brown steak in 2 tablespoons oil in skillet; drain. Add remaining 2 tablespoons oil and next 4 ingredients. Sauté for 5 minutes. Stir in flour. Cook for 1 minute. Stir in bouillon dissolved in boiling water and remaining ingredients. Bring to a boil. Simmer, covered, for 2 minutes. Add steak. Cook until steak is tender, stirring occasionally. Serve over rice or noodles. Yield: 8 servings.

Approx per serving: Cal 391, Prot 16.8 gr,
T Fat 31.9 gr, Chol 58.0 mg, Carbo 9.1 gr,
Sod 533.0 mg, Potas 648.0 mg.

Abby Brown
Pine Bluff, AR

Swiss Steak

1½ lb. ¾-inch thick boneless
* round steak*
¼ c. flour
1 tsp. salt
1 c. water
¾ tsp. Worcestershire sauce
2 tbsp. chopped onion
1 16-oz. can tomatoes
½ c. chopped celery
½ c. chopped carrots

Trim fat from steak; cut steak into 6 pieces. Render fat in skillet; discard, reserving drippings. Pound mixture of flour and salt into steak; reserve remaining flour mixture. Brown steak in drippings in skillet. Place in 1½-quart baking dish. Blend reserved flour mixture into drippings in skillet. Stir in water gradually. Cook until slightly thickened, stirring constantly. Add remaining ingredients. Pour over steak. Bake, covered, at 325 degrees for 1 hour. Bake, uncovered, for 1 hour longer. Yield: 6 servings.

Approx per serving: Cal 265, Prot 24.0 gr, T Fat 14.3 gr, Chol 77.4 mg, Carbo 9.0 gr, Sod 557.3 mg, Potas 608.2 mg.

Wanda Carson
Jackson, MS

Teriyaki

1 lg. onion, chopped
1 lb. round steak, thinly sliced
1 tsp. garlic salt
2 tbsp. oil
2 tsp. ground cloves
1 12-oz. can tomato juice
½ c. dark corn syrup
Pinch of basil

Sauté onion and beef with garlic salt in oil in skillet until brown. Stir in remaining ingredients. Cook, covered, over low heat for 2 hours or until tender, adding a small amount of tomato juice if necessary. Yield: 6 servings.

Approx per serving: Cal 305, Prot 15.4 gr, T Fat 16.3 gr, Chol 51.4 mg, Carbo 24.3 gr, Sod 548.0 mg, Potas 395.7 mg.

Veronica Bradley
McEwen, TN

Yankee Stir-Fry

1 lb. round steak, thinly sliced
3 tbsp. soy sauce
1 tbsp. Sherry
½ tsp. ginger
1 4-oz. can chopped mushrooms
1 tbsp. oil
1½ c. diagonally sliced carrots
1 tbsp. oil
1½ c. thinly sliced cauliflower
1 10-oz. package frozen peas
4 to 6 green onions, cut into ½-in. pieces
1 tbsp. oil
½ c. cold water
2 tbsp. cornstarch
½ tsp. sugar

Combine steak with soy sauce, Sherry and ginger in bowl; mix well. Marinate for 15 to 30 minutes. Drain mushrooms, reserving liquid. Heat 1 tablespoon oil to 375 degrees in wok. Add carrots. Stir-fry for 2 minutes. Add mushroom liquid. Steam, covered, for 4 minutes. Remove carrots with slotted spoon; drain wok. Heat 1 tablespoon oil in wok. Add cauliflower and peas. Stir-fry for 2 minutes; push to one side. Add green onions and mushrooms. Stir-fry for 2 minutes; push to one side. Add remaining 1 tablespoon oil and beef mixture. Stir-fry for 2 to 3 minutes or until tender. Stir in carrots. Combine water, cornstarch and sugar in small bowl. Stir into wok. Cook until thickened, stirring constantly. Yield: 6 servings.

Approx per serving: Cal 310, Prot 19.3 gr, T Fat 15.3 gr, Chol 49.2 mg, Carbo 17.5 gr, Sod 774.4 mg, Potas 481.1 mg.

Robin Bartoletti
Riverside Jr.-Sr. H.S., Dickson City, PA

Breaded Veal Cutlets

1 lb. thin veal cutlets
1 tsp. salt
½ tsp. paprika
½ c. frozen egg substitute, thawed
1 c. fine dry bread crumbs
1 tbsp. minced parsley
Corn oil for frying

Sprinkle veal cutlets with mixture of salt and paprika. Pound with meat mallet to flatten. Dip into egg substitute. Coat with mixture of bread crumbs and parsley. Let dry on wire rack for 15 minutes. Heat ¼ inch corn oil to 375 degrees in skillet. Brown cutlets for 3 to 4 minutes on each side; drain. Place on serving plate. Garnish with lemon slices. Yield: 4 servings.
Note: Nutritional information does not include corn oil for frying.

Approx per serving: Cal 288, Prot 24.5 gr, T Fat 12.0 gr, Chol 83.1 mg, Carbo 18.3 gr, Sod 769.3 mg, Potas 277.6 mg.

Theresa Babonis
Pittston Area Sr. H.S., Dupont, PA

American Chop Suey

1 lb. ground round
2 tbsp. shortening
1 lg. onion, chopped
1 green pepper, chopped
¾ c. rice
1 c. chopped celery
2 c. chopped tomatoes
1 tsp. salt

Brown ground round in shortening in saucepan. Add remaining ingredients; mix well. Cook, covered, over high heat until steaming; remove from heat. Let stand for 1 hour. Yield: 8 servings.

Approx per serving: Cal 215, Prot 13.8 gr, T Fat 9.4 gr, Chol 38.0 mg, Carbo 18.3 gr, Sod 331.0 mg, Potas 382.0 mg.

Alice King
Danville, KY

Barley Hot Dish

1½ lb. ground beef
1 c. chopped onion
½ c. chopped celery
4 c. tomato juice
¾ c. barley
1 lb. peas
1 lb. carrots, chopped
1 can cream of mushroom soup
2½ tsp. salt
¼ tsp. pepper
1 bay leaf
2½ c. water

Brown ground beef with onion and celery in skillet, stirring frequently; drain. Stir in remaining ingredients. Bring to a boil. Pour into large baking dish. Bake, covered, at 375 degrees for 1¼ hours. Yield: 20 servings.

Approx per serving: Cal 170, Prot 8.8 gr, T Fat 8.7 gr, Chol 24.3 mg, Carbo 14.6 gr, Sod 519.5 mg, Potas 359.4 mg.

Bessie Hockman
Cairo, IL

Deviled Hamburgers

1 lb. ground round
1 tbsp. grated onion
1 clove of garlic, minced
½ tsp. mustard
1 tsp. tomato paste
½ tsp. minced dill pickle
½ tsp. Worcestershire sauce

Combine all ingredients in bowl; mix well. Shape into 6 patties. Place on rack in broiler pan. Broil to desired degree of doneness. Yield: 6 servings.

Approx per serving: Cal 128, Prot 15.6 gr, T Fat 6.4 gr, Chol 53.3 mg, Carbo 0.7 gr, Sod 56.2 mg, Potas 190.6 mg.

Gladys Stuart
Miami, OH

Crock • Pot Texas Hash

2 lb. ground beef
2 med. onions, chopped
2 green peppers, chopped
2 16-oz. cans tomatoes
1 c. rice
2 tsp. Worcestershire sauce
2½ tsp. chili powder
2½ tsp. salt

Brown ground beef in skillet, stirring until crumbly; drain. Place in Crock • Pot. Add remaining ingredients; mix well. Cook, covered, on Low for 6 to 8 hours, adding water if necessary. Yield: 8 servings.

Approx per serving: Cal 360, Prot 23.3 gr, T Fat 17.0 gr, Chol 77.2 mg, Carbo 28.1 mg, Sod 891.3 mg, Potas 594.0 mg.

Easter Moore
Abbeville, SC

Beef and Bean Casserole

1 lb. ground beef
1 onion, chopped
1 16-oz. can French-style green beans
1 16-oz. package frozen Tater Tots
1 can cream of chicken soup
1 soup can milk

Brown ground beef with onion in skillet, stirring frequently; drain. Spread in 9x13-inch baking dish. Drain green beans. Layer beans and Tater Tots over ground beef. Mix soup and milk in small bowl. Pour over layers. Bake at 350 degrees for 1 hour and 15 minutes. Yield: 6 servings.

Approx per serving: Cal 306, Prot 18.1 gr, T Fat 15.2 gr, Chol 61.9 mg, Carbo 24.4 gr, Sod 640.0 mg, Potas 492.8 mg.

Ina Pack
Christiansburg H.S., Radford, VA

Cabbage Casserole

1 lb. ground beef
1 med. onion, chopped
½ c. chopped green pepper
1 can mushroom soup
½ c. water
1 med. head cabbage, shredded
1½ c. sliced carrots
1 tbsp. butter, melted
2 tbsp. flour
2 c. milk
6 oz. Velveeta cheese, chopped

Brown ground beef in skillet, stirring until crumbly; drain. Sprinkle with seasoned salt to taste. Add onion and green pepper. Cook for 5 to 10 minutes or until vegetables are tender. Stir in mushroom soup. Simmer for 10 to 15 minutes, adding water as necessary to make of desired consistency. Combine cabbage and carrots with a small amount of water in saucepan. Steam, covered, until tender-crisp. Blend butter and flour in saucepan. Stir in milk gradually. Cook until thickened, stirring constantly. Stir in cheese until melted. Alternate layers of ground beef mixture, vegetables and cheese sauce in baking dish until all ingredients are used. Broil until bubbly. Yield: 4 servings.

Approx per serving: Cal 639, Prot 37.2 gr, T Fat 40.3 gr, Chol 139.1 mg, Carbo 33.3 gr, Sod 1467.4 mg, Potas 1104.9 mg.

Lynette Ann Gossen
Crowley H.S., Lafayette, LA

Family Favorite Beef in Casserole

1 lb. ground beef
2 tsp. sugar
2 8-oz. cans Spanish-style tomato sauce
2 tsp. garlic salt
5 oz. egg noodles

8 oz. cream cheese, chopped
1 c. sour cream
1 c. grated Cheddar cheese

Brown ground beef in skillet, stirring until crumbly; drain. Add sugar, tomato sauce and garlic salt; mix well. Cook noodles according to package directions; drain. Mix with cream cheese and sour cream. Layer beef mixture, noodle mixture and Cheddar cheese ½ at a time in greased 1½-quart baking dish. Bake at 375 degrees for 35 minutes. Yield: 6 servings.

Approx per serving: Cal 580, Prot 26.4 gr, T Fat 40.7 gr, Chol 150.9 mg, Carbo 27.8 gr, Sod 1428.0 mg, Potas 587.1 mg.

Carol Brown
Fort Osage Jr. H.S., Independence, MO

Fiesta Casserole

1½ lb. ground beef
¼ tsp. Tabasco sauce
Pepper to taste
1 med. onion, chopped
½ green pepper, chopped
¼ tsp. garlic powder
½ tsp. chili powder
½ tsp. salt
1½ c. macaroni, cooked
½ c. Parmesan cheese
1 can tomato soup
1 soup can milk
½ c. bread crumbs

Brown ground beef in skillet, stirring until crumbly; drain. Stir in next 7 ingredients. Layer macaroni, ground beef mixture and cheese alternately in greased 2-quart casserole until all ingredients are used. Mix soup and milk in bowl. Pour over casserole. Sprinkle bread crumbs over top. Bake at 350 degrees for 30 to 40 minutes or until heated through. Yield: 8 servings.

Approx per serving: Cal 407, Prot 23.2 gr, T Fat 22.6 gr, Chol 70.7 mg, Carbo 26.5 gr, Sod 586.3 mg, Potas 464.0 mg.

Chloe Oschenbein
Sacramento, CA

Lean Chili

2 c. sliced mushrooms
1 c. chopped celery
1 c. chopped onion
2 tbsp. margarine
1 lb. lean ground beef
2 tbsp. chili powder
¼ tsp. salt
½ tsp. pepper
2   28-oz. cans tomatoes
1 tbsp. sugar
1 tsp. Worcestershire sauce
¼ c. water
2 c. kidney beans

Sauté mushrooms, celery and onion in margarine in skillet until onion is tender. Add ground beef and about ¼ of the mixture of chili powder, salt and pepper. Cook until ground beef is brown, stirring frequently. Stir in remaining seasoning mixture, tomatoes, sugar, Worcestershire sauce and water. Simmer, covered, for 15 minutes. Add kidney beans. Simmer for 10 minutes longer. Spoon into serving bowls. Yield: 6 servings.

Approx per serving: Cal 292, Prot 23.0 gr,
T Fat 11.0 gr, Chol 53.3 mg, Carbo 26.1 gr,
Sod 387.9 mg, Potas 902.6 mg.

Dallas Tudor
Casper, WY

Double-Threat Chili

8 oz. ground round
8 oz. ground fresh turkey
1   28-oz. can tomatoes
1   8-oz. can tomato sauce
¾ c. chopped carrots
¾ c. chopped onion
½ c. chopped green pepper
1 tsp. minced garlic
1 tbsp. chili powder
1 bay leaf
1 c. drained red kidney beans
2 c. cooked long grain rice

Brown ground beef and ground turkey in skillet, stirring frequently; drain. Add tomatoes, tomato sauce, carrots, onion, green pepper and seasonings. Simmer, partially covered, for 1 hour. Stir in beans. Simmer for 15 minutes longer. Remove bay leaf. Spoon over hot rice in serving bowls. Yield: 4 servings.

Approx per serving: Cal 366, Prot 23.6 gr,
T Fat 6.4 gr, Chol 46.1 mg, Carbo 54.9 gr,
Sod 1047.3 mg, Potas 1183.1 mg.

Sue Fayar
Charlotte, NC

Super Stuffed Peppers

8 oz. lean ground beef
1 lg. onion, chopped
6 med. tomatoes, chopped
8 lg. green peppers, seeded
3 c. grated Monterey Jack cheese
3 c. kidney beans
2 tbsp. chopped parsley
1 clove of garlic, minced
¼ tsp. each basil, oregano,
   salt and pepper

Brown ground beef in skillet, stirring frequently; drain. Sauté onion and tomatoes in skillet; drain. Parboil green peppers in salted water in saucepan for 3 minutes; drain. Mix ground beef, tomato mixture and remaining ingredients in bowl. Spoon into green peppers. Arrange in baking pan. Bake, covered, at 325 degrees for 30 minutes. Yield: 8 servings.

Approx per serving: Cal 346, Prot 24.1 gr,
T Fat 16.8 gr, Chol 61.9 mg, Carbo 26.4 gr,
Sod 394.9 mg, Potas 769.8 mg.

Ophelia Sansing
Morton, MS

Stuffed Green Peppers

3 lg. green peppers
½ lb. ground beef
1   8-oz. can tomato sauce
½ c. cracker crumbs
1 tbsp. chopped onion
1 tsp. salt
¼ tsp. pepper

Cut thin slice from stem end of peppers; discard seed. Cook in boiling salted water to cover in saucepan for 5 minutes; drain. Combine remaining ingredients in bowl; mix well. Spoon lightly into peppers. Place in small baking dish.

Bake, covered, at 350 degrees for 45 minutes. Bake, uncovered, for 15 minutes longer.
Yield: 3 servings.

Approx per serving: Cal 297, Prot 16.6 gr, T Fat 17.7 gr, Chol 51.4 mg, Carbo 18.1 gr, Sod 1330.7 mg, Potas 656.9 mg.

Linda Mallory
Roanoke, VA

Hobo Dinner

   1 lb. ground beef
   ¼ c. evaporated milk
   1 tsp. salt
   ½ tsp. pepper
   6 lg. potatoes, thinly sliced
   2 lg. onions, sliced
   1 can cream of mushroom soup
   ¼ lb. sharp cheese, grated

Press ground beef into bottom of 2½-quart casserole. Sprinkle with evaporated milk, salt and pepper. Layer potatoes, onions, soup and cheese over top. Bake, covered, at 350 degrees for 1 hour or until potatoes are tender.
Yield: 8 servings.

Approx per serving: Cal 331, Prot 15.2 gr, T Fat 16.1 gr, Chol 44.9 mg, Carbo 31.7 gr, Sod 622.1 mg, Potas 856.6 mg.

Sarah Crossway
Mobile, AL

Lasagna

   1 lb. ground beef
   1 clove of garlic, minced
   1 tbsp. oil
   2 8-oz. cans tomato sauce
   1 6-oz. can tomato paste
   1 6-oz. can water
   1 tsp. chopped parsley
   ¼ tsp. oregano
   1 tsp. salt
   ¼ tsp. pepper
   1 16-oz. package lasagna noodles
   1 16-oz. carton ricotta cheese
   ¼ c. Parmesan cheese
   2 c. shredded American cheese
   3 c. shredded mozzarella cheese

Brown ground beef with garlic in oil in saucepan, stirring frequently; drain. Add tomato sauce, tomato paste, water, parsley and seasonings; mix well. Simmer for 30 minutes. Cook noodles according to package directions; drain. Spread ¼ of the meat sauce in 9x13-inch baking dish. Layer noodles, cheeses and remaining meat sauce ⅓ at a time in prepared dish. Bake, covered, at 350 degrees for 40 minutes. Bake, uncovered, for 5 minutes longer. Let stand for several minutes. Cut into squares to serve.
Yield: 6 servings.

Approx per serving: Cal 1038, Prot 67.9 gr, T Fat 52.3 gr, Chol 260.6 mg, Carbo 71.9 gr, Sod 1883.8 mg, Potas 1131.1 mg.

Deborah Lanuti
Valley View H.S., Scranton, PA

Three-Cheese Lasagna

   8 oz. lean ground beef
   ½ tsp. basil
   1 tsp. oregano
   8 oz. tomato paste
   6 tbsp. grated mozzarella cheese
   ⅔ c. cottage cheese
   ¼ c. Parmesan cheese
   1 c. low-fat milk
   ¼ c. flour
   2 tsp. baking powder
   2 eggs
   Pinch of pepper
   6 tbsp. grated mozzarella cheese

Brown ground beef in skillet, stirring frequently; drain. Add basil, oregano, tomato paste and 6 tablespoons mozzarella cheese; mix well. Layer cottage cheese, Parmesan cheese and ground beef mixture in 8-inch square baking pan sprayed with nonstick cooking spray. Combine milk, flour, baking powder, eggs, pepper and salt to taste in blender container. Process for 30 seconds. Pour over layers. Sprinkle with 6 tablespoons mozzarella cheese. Bake at 400 degrees for 30 to 35 minutes or until golden.
Yield: 4 servings.

Approx per serving: Cal 496, Prot 30.4 gr, T Fat 32.3 gr, Chol 223.6 mg, Carbo 21.3 gr, Sod 506.2 mg, Potas 787.6 mg.

Susan Miller
Los Angeles, CA

No-Noodle Lasagna

8 oz. lean ground beef
½ c. chopped onion
1   15-oz. can tomato sauce
½ tsp. each oregano, basil and salt
⅛ tsp. pepper
4 med. zucchini
1   8-oz. carton cottage cheese
1 egg
2 tbsp. flour
1 c. shredded low-fat mozzarella cheese

Brown ground beef with onion in skillet, stirring frequently; drain. Stir in tomato sauce, oregano, basil, salt and pepper. Simmer for 5 minutes. Slice zucchini lengthwise into ¼-inch slices. Combine cottage cheese and egg in bowl; mix well. Layer half the zucchini, half the flour, all the cottage cheese mixture, half the ground beef mixture, remaining zucchini and flour, mozzarella cheese and remaining ground beef mixture in 8x12-inch baking dish. Bake at 375 degrees for 40 minutes. Let stand for 10 minutes before serving. Yield: 6 servings.

Approx per serving: Cal 251, Prot 22.2 gr,
T Fat 10.8 gr, Chol 93.1 mg, Carbo 17.6 gr,
Sod 818.4 mg, Potas 829.9 mg.

Anne Craighorn
Billings, MT

Juicy Meat Loaf

2 lb. lean ground beef
½ c. oats
1 onion, chopped
½ c. tomato juice
2 tsp. salt
Pepper to taste
3 slices onion
1 c. tomato juice

Combine first 6 ingredients in bowl; mix well. Shape into loaf in baking dish. Top with onion slices. Pour 1 cup tomato juice over loaf. Bake at 325 degrees for 1½ hours, basting occasionally with pan juices. Skim juices before serving. Yield: 8 servings.

Approx per serving: Cal 273, Prot 21.3 gr,
T Fat 17.0 gr, Chol 76.6 mg, Carbo 8.1 gr,
Sod 675.7 mg, Potas 392.0 mg.

Wendy Hetz
Ft. Smith, AR

Meat Loaf

3 lb. lean ground beef
2 c. fresh bread crumbs
½ c. chopped onion
¼ c. chopped green pepper
¼ c. minced parsley
½ c. catsup
⅓ c. water
2 tbsp. Worcestershire sauce
1 clove of garlic, minced
½ c. frozen egg substitute, thawed
½ tsp. oregano
1 tsp. salt

Combine all ingredients in bowl; mix well. Pack into ungreased 5x9-inch loaf pan. Bake at 350 degrees for 1½ hours; drain. Invert onto serving platter. Let stand for several minutes before slicing. Yield: 12 servings.
Note: Nutritional information does not include egg substitute.

Approx per serving: Cal 228, Prot 24.5 gr,
T Fat 10.0 gr, Chol 80.2 mg, Carbo 8.6 gr,
Sod 412.4 mg, Potas 346.2 mg.

Theresa Babonis
Pittston Area Sr. H.S., Dupont, PA

All-Natural Pizza

1¼ c. unbleached flour
¾ c. oat flour
1 tsp. baking powder
1 tsp. salt
⅔ c. milk
¼ c. oil
1   15-oz. can tomato sauce
1 tsp. honey
1 tsp. onion flakes
1 tbsp. Italian seasoning
½ tsp. oregano
1 tsp. salt
3 c. cooked ground beef
2 c. grated mozzarella cheese
½ c. Parmesan cheese

Combine unbleached flour, oat flour, baking powder and 1 teaspoon salt in mixing bowl. Add milk and oil. Stir until dough forms ball. Knead several times on floured surface. Pat into greased 14-inch pizza pan. Bake at 425 degrees for 12 to 14 minutes or until light brown. Combine tomato sauce, honey, onion flakes, Italian seasoning, oregano and 1 teaspoon salt in bowl; mix well. Spread on crust. Sprinkle with ground beef and cheeses. Bake for 10 to 15 minutes longer or until bubbly. Yield: 6 servings.

Approx per serving: Cal 608, Prot 41.3 gr, T Fat 32.5 gr, Chol 126.4 mg, Carbo 36.8 gr, Sod 1501.5 mg, Potas 712.9 mg.

Zeta Davidson
Oak Park H.S., Kansas City, MO

No-Crust Pizza

1 2-oz. can chopped mushrooms
1 egg, slightly beaten
1 c. soft bread crumbs
½ tsp. each oregano, salt
Pinch of pepper
1 lb. lean ground beef
2 1-oz. slices mozzarella cheese
½ c. pizza sauce
¼ c. chopped onion
2 tbsp. chopped green pepper

Drain mushrooms, reserving liquid. Add enough water to reserved liquid to measure ⅓ cup. Combine with egg, bread crumbs, oregano, salt and pepper in bowl; mix well. Let stand for 5 minutes. Add ground beef; mix well. Pat into 9-inch pie plate, forming crust. Cut cheese into 8 triangles. Arrange half the cheese over ground beef. Layer pizza sauce, mushrooms, onion and green pepper over cheese. Bake at 350 degrees for 45 minutes. Top with remaining cheese. Bake for 5 minutes longer. Yield: 8 servings.

Approx per serving: Cal 147, Prot 14.5 gr, T Fat 8.2 gr, Chol 76.5 mg, Carbo 2.7 gr, Sod 230.1 mg, Potas 225.6 mg.

Judy Meek
Marshall Jr. H.S., Wichita, KS

Whole Wheat Pizza

½ pkg. dry yeast
1 tsp. honey
2 tbsp. warm water
½ c. stirred whole wheat flour
1 tbsp. oil
½ tsp. salt
6 tbsp. warm water
1 egg
1½ c. stirred whole wheat flour

2 8-oz. cans tomato sauce
2 tbsp. brown sugar
¼ tsp. oregano
2 lb. ground beef
1 green pepper, chopped
2 4-oz. cans mushrooms, drained
16 1-oz. slices American cheese

Dissolve yeast with honey in 2 tablespoons warm water. Combine ½ cup flour, oil, salt and 6 tablespoons warm water in bowl. Add yeast mixture and egg; beat until smooth. Stir in remaining 1½ cups flour to make a soft dough. Place in greased bowl, turning to grease surface. Let rise for 30 minutes to 1 hour or until doubled in bulk. Knead for several minutes. Divide into 2 portions. Pat over bottoms of 2 oiled 9x13-inch baking dishes. Spread mixture of tomato sauce, brown sugar and oregano evenly over dough. Brown ground beef in skillet, stirring until crumbly; drain. Sprinkle ground beef, green pepper and mushrooms over sauce. Top with cheese slices. Bake at 450 degrees for 15 minutes. Reduce temperature to 375 degrees. Bake until light brown. Yield: 6 servings.

Approx per serving: Cal 876, Prot 50.4 gr, T Fat 54.4 gr, Chol 199.4 mg, Carbo 47.4 gr, Sod 1931.0 mg, Potas 1128.4 mg.

Sharie Mueller
Jefferson Co. North H.S., Oskaloosa, KS

Summer Sausage

2 lb. ground beef
¾ c. water
1 tbsp. mustard seed
1 tbsp. liquid smoke
1 tsp. pepper
¼ tsp. onion salt
½ tsp. garlic salt
2 tbsp. Morton Tender-Quick salt

Combine all ingredients in bowl; mix well. Shape into 2 rolls; wrap in foil. Chill for 24 hours. Place on rack in baking pan. Bake at 350 degrees for 1 hour. Chill in refrigerator. Slice and serve with crackers. Yield: 12 servings.

Approx per serving: Cal 155, Prot 13.1 gr, T Fat 11.0 gr, Chol 51.1 mg, Carbo 0.0 gr, Sod 1186.9 mg, Potas 147.4 mg.

Linda Wright
Idabel H.S., Foreman, AR

Burger Rolls

1 c. grated Cheddar cheese
1 c. coarse cracker crumbs
1 c. milk
2 tbsp. catsup
¼ c. packed brown sugar
½ med. onion, chopped
½ tsp. celery salt
1 tsp. salt
¼ tsp. pepper
2 lb. lean ground beef
12 slices bacon

Combine first 9 ingredients in bowl. Add ground beef; mix well. Shape into 12 rolls. Wrap bacon slice around each roll; secure with wooden pick. Place on rack in broiler pan. Broil 5 inches from heat source for 10 minutes; turn rolls. Broil for 8 to 10 minutes longer or until bacon is crisp. Yield: 12 servings.
Note: Rolls may be prepared, wrapped tightly and stored in freezer. Thaw before cooking.

Approx per serving: Cal 267, Prot 21.2 gr,
T Fat 14.8 gr, Chol 72.1 mg, Carbo 11.3 gr,
Sod 548.1 mg, Potas 271.8 mg.

Diane Zook
Larned H.S., Larned, KS

Speedy Springtime Supper

1 lb. ground beef
1 sm. clove of garlic, minced
⅛ tsp. cumin
⅛ tsp. nutmeg
1 tsp. salt
Pinch of pepper
1 tbsp. flour
½ c. water
1   10-oz. package frozen peas, blanched
1 med. green pepper, cut into thin strips

Combine ground beef, garlic, cumin, nutmeg, salt and pepper in bowl; mix lightly. Shape into 12 meatballs. Brown on all sides in skillet; pour off drippings. Mix flour and water in small bowl. Add to meatballs. Stir in peas and green pepper. Cook, covered, for 5 to 7 minutes or until vegetables are tender-crisp. Yield: 4 servings.

Photograph for this recipe on page 39.

Victory Garden Skillet Dinner

1 lb. ground beef
2½ c. water
1 c. rice
1 sm. onion, chopped
4 beef bouillon cubes
½ tsp. each dry mustard, MSG
4 slices American cheese, chopped
1 med. green pepper, chopped
1   15-oz. can tomatoes
2 slices American cheese,
*   cut into triangles*

Brown ground beef in 10-inch skillet, stirring until crumbly; drain. Add next 6 ingredients; mix well. Bring to a boil; reduce heat. Simmer, covered, for 20 minutes. Add chopped cheese, green pepper and tomatoes; mix well. Arrange cheese triangles on top; remove from heat. Let stand, covered, for 5 minutes or until liquid is absorbed. Yield: 6 servings.

Approx per serving: Cal 388, Prot 22.7 gr,
T Fat 18.2 gr, Chol 73.5 mg, Carbo 32.0 gr,
Sod 1300.6 mg, Potas 455.7 mg.

Marilyn Mancewicz
Ottawa Hills H.S., Grand Rapids, MI

Brown Hamburger Stew

1 lb. lean ground beef
¼ c. flour
2 tsp. salt
¼ tsp. pepper
2 tbsp. oil
1½ c. water
3 carrots, sliced
3 potatoes, chopped
2 onions, sliced
2 c. tomato juice

Combine ground beef, flour, salt and pepper in bowl; mix well. Brown in oil in saucepan, stirring until crumbly; drain. Add water and carrots. Simmer, covered, for 10 minutes. Add potatoes and onions. Simmer for 10 minutes or until tender-crisp. Add tomato juice. Simmer until heated through. Yield: 6 servings.

Approx per serving: Cal 306, Prot 20.0 gr,
T Fat 11.3 gr, Chol 53.3 mg, Carbo 31.9 gr,
Sod 936.1 mg, Potas 956.2 mg.

Lynn G. Holmes
Arrowhead Mid. Sch., Kansas City, KS

Ham and Broccoli Bake

> 1   10-oz. package frozen
>   chopped broccoli
> 12 slices white bread
> 1 c. shredded Cheddar cheese
> 2 c. chopped cooked ham
> 2 tsp. chopped onion
> 6 eggs, beaten
> 3½ c. milk
> ½ tsp. salt
> ¼ to ½ tsp. dry mustard

Cook broccoli according to package directions until partially cooked; drain. Trim and discard crusts from bread. Cut each slice with doughnut cutter. Place bread scraps in greased 9x13-inch baking dish. Layer cheese, broccoli and ham in prepared dish. Sprinkle with onion. Arrange bread doughnuts over top. Beat eggs, milk, salt and dry mustard in bowl. Pour over layers. Refrigerate, covered, for 6 hours or longer. Bake at 325 degrees for 1 hour. Let stand for 10 minutes. Cut into 12 squares. Yield: 12 servings.

Approx per serving: Cal 276, Prot 16.2 gr,
T Fat 14.5 gr, Chol 167.3 mg, Carbo 19.5 gr,
Sod 543.3 mg, Potas 284.9 mg.

Freda Smith
Fort Wayne, IN

Impossible Ham and Swiss Pie

> 2 c. chopped cooked ham
> 1 c. shredded Swiss cheese
> ⅓ c. chopped green onions
> 4 eggs
> 2 c. milk
> 1 c. buttermilk baking mix
> ¼ tsp. salt
> ⅛ tsp. pepper

Sprinkle ham, cheese and green onions in greased 10-inch pie plate. Combine remaining ingredients in bowl; beat until smooth. Pour over ham. Bake at 400 degrees for 35 to 40 minutes or until golden brown and knife inserted in center comes out clean. Cool for 5 minutes before serving. Yield: 6 servings.

Approx per serving: Cal 398, Prot 23.7 gr,
T Fat 24.8 gr, Chol 240.4 mg, Carbo 18.6 gr,
Sod 913.9 mg, Potas 317.5 mg.

Pat Dorman
Eddy County, NM

Cauliflower-Ham Casserole

> 2 tbsp. melted butter
> 2 tbsp. flour
> 1 c. milk
> 4 slices sharp American cheese
> ½ c. sour cream
> 1 med. head cauliflower, chopped,
>   cooked
> 2 c. chopped cooked ham
> 1   3-oz. can mushrooms, drained
> ½ c. dry bread crumbs

Blend butter and flour in saucepan. Stir in milk gradually. Cook until thickened, stirring constantly. Stir in cheese until melted. Remove from heat. Stir in sour cream. Pour over mixture of drained cauliflower, ham and mushrooms in 1½-quart casserole. Sprinkle crumbs over top. Bake at 350 degrees for 30 minutes. Yield: 6 servings.

Approx per serving: Cal 380, Prot 20.9 gr,
T Fat 24.9 gr, Chol 81.5 mg, Carbo 20.0 gr,
Sod 807.8 mg, Potas 672.7 mg.

Sheila Trenton
Baltimore, MD

Special Fried Rice

> ½ c. chopped onion
> 3 tbsp. oil
> ½ c. chopped mushrooms
> ½ c. bean sprouts
> 1½ c. sliced celery
> 1½ c. chopped cooked ham
> 6 c. cooked rice
> 1 tbsp. soy sauce
> 2 eggs, scrambled
> ½ c. chopped green onion

Stir-fry onion in hot oil in skillet until brown. Add mushrooms, bean sprouts and celery. Stir-fry until tender. Add ham, rice and salt and pepper to taste and enough soy sauce to make of desired color; mix well. Spoon into serving bowl. Cut eggs into strips. Arrange over top. Sprinkle with green onion. Yield: 6 servings.

Approx per serving: Cal 436, Prot 15.2 gr,
T Fat 16.9 gr, Chol 115.4 mg, Carbo 54.5 gr,
Sod 1424.1 mg, Potas 380.5 mg.

Nettie Boelker
Larchmont, NY

Cheese and Ham Delight

½ c. finely chopped onion
1 tbsp. butter
2 c. chopped cooked ham
1 c. shredded sharp American cheese
3 eggs, beaten
⅔ c. fine cracker crumbs
1½ c. milk

Sauté onion in butter in skillet. Add remaining ingredients; mix well. Spoon into 6x10-inch baking dish. Bake at 350 degrees for 50 minutes. Yield: 6 servings.

Approx per serving: Cal 287, Prot 17.0 gr, T Fat 19.3 gr, Chol 185.8 mg, Carbo 10.4 gr, Sod 596.4 mg, Potas 272.4 mg.

Hassie H. Green
Goshen H.S., Goshen, AL

Ham Turnovers

1 c. minced baked ham
⅓ c. chopped celery
¼ c. chopped walnuts
1 tsp. minced onion
3 tbsp. sweet pickle relish
½ tsp. mustard
¼ c. salad dressing
¼ c. butter, softened
12 slices whole wheat bread,
    crusts trimmed
½ c. Cheez Whiz

Combine ham, celery, walnuts, onion, relish, mustard and salad dressing in bowl; mix well. Butter bread lightly; spread with Cheez Whiz. Place about 1 tablespoon ham mixture on center of each. Fold over to form triangle, enclosing filling; secure with toothpick. Store in tightly covered container in refrigerator for several hours if desired. Place on baking sheet. Bake at 450 degrees for 5 to 6 minutes or until toasted. Yield: 12 servings.

Approx per serving: Cal 188, Prot 6.5 gr, T Fat 12.0 gr, Chol 28.9 mg, Carbo 14.8 gr, Sod 404.0 mg, Potas 138.6 mg.

Cindi McGuffey
Prescott, AR

Quick Quiche

3 eggs
½ c. melted butter
1½ c. milk
½ c. buttermilk baking mix
½ tsp. salt
⅛ tsp. pepper
1 c. grated Swiss cheese
1 c. chopped ham
1 c. sliced onion

Combine first 6 ingredients in blender container. Process until smooth. Pour into bowl. Stir in cheese, ham and onion. Pour into greased pie plate. Bake at 350 degrees for 45 minutes or until set. Yield: 6 servings.
Note: May substitute meat, cheese and vegetables of your choice.

Approx per serving: Cal 366, Prot 15.5 gr, T Fat 28.2 gr, Chol 213.6 mg, Carbo 12.9 gr, Sod 820.9 mg, Potas 236.9 mg.

Allison Wilson
San Mateo, CA

Verenike Casserole

1½ to 2 c. dry curd cottage cheese
2 eggs
1   8-oz. package wide noodles, cooked
2 c. sour cream
1½ c. chopped ham
1 c. half and half

Combine cottage cheese and eggs in bowl; mix well. Season with salt and pepper to taste. Layer noodles, cottage cheese mixture and sour cream in 9x13-inch baking dish. Brown ham in skillet. Add half and half, stirring to deglaze skillet. Pour over layers. Bake at 350 degrees for 30 to 40 minutes or until bubbly. Yield: 4 servings.

Approx per serving: Cal 802, Prot 39.4 gr, T Fat 48.5 gr, Chol 307.9 mg, Carbo 50.7 gr, Sod 725.0 mg, Potas 527.6 mg.

Valerie Wall
Wichita Northwest H.S., Wichita, KS

Sweet and Sour Pork

1½ lb. pork, cut into chunks
2 c. water
1 tbsp. soy sauce
1 c. tomatoes, chopped
1 green pepper, chopped
½ c. chopped onion
⅔ c. pineapple chunks
½ c. vinegar
¼ c. each sugar, packed brown sugar
¼ c. cornstarch
½ c. pineapple juice
½ c. water

Cook pork in 2 cups water in saucepan until tender. Add soy sauce, vegetables and pineapple; mix well. Combine remaining ingredients in bowl; mix well. Stir into pork mixture. Bring to a boil, stirring constantly. Serve over rice. Yield: 6 servings.

Approx per serving: Cal 483, Prot 21.0 gr,
T Fat 28.6 gr, Chol 70.3 mg, Carbo 35.8 gr,
Sod 514.0 mg, Potas 586.7 mg.

Mollie Geise
Northumberland, PA

Easy and Delicious Pork Stir-Fry

2 tbsp. oil
¾ lb. pork tenderloin, cut into
*   bite-sized pieces*
1 onion, cut into med. strips
¾ c. water
1 env. onion gravy mix
2 tbsp. each vinegar, soy sauce
2 tbsp. sugar

Heat oil in wok. Add pork and onion. Stir-fry until pork is cooked through; drain. Combine remaining ingredients in bowl; mix well. Add to pork. Cook until thickened, stirring constantly. Serve over hot cooked rice. Yield: 4 servings. Note: Nutritional information does not include onion gravy mix.

Approx per serving: Cal 318, Prot 15.3 gr,
T Fat 23.4 gr, Chol 51.4 mg, Carbo 11.5 gr,
Sod 698.3 mg, Potas 265.8 mg.

Karen R. Collins
Greensville Co. H.S., Emperia, VA

Cheese-Sausage Soufflé

1 lb. link sausages
6 eggs, beaten
6 slices bread, cubed
2 c. milk
1 c. shredded Cheddar cheese
1 tsp. dry mustard
1 tsp. salt

Brown sausages in skillet; drain and slice. Combine with remaining ingredients in bowl; mix well. Spoon into greased 9x13-inch baking dish. Chill overnight. Bake at 350 degrees for 45 minutes. Yield: 8 servings.

Approx per serving: Cal 341, Prot 17.2 gr,
T Fat 23.5 gr, Chol 236.5 mg, Carbo 14.4 gr,
Sod 804.6 mg, Potas 241.9 mg.

Leslie K. Donnell
Talawanda H.S., Oxford, OH

Sauerkraut and Sausage

1 lb. Polish sausage,
*   cut into 3-in. pieces*
1   16-oz. jar sauerkraut
1 apple, sliced
¼ c. packed brown sugar
1 tsp. caraway seed
1 tsp. paprika

Cook sausage in large skillet for 10 minutes or until brown; drain. Drain sauerkraut, reserving ½ cup liquid. Rinse sauerkraut. Add sauerkraut to sausage with apple and brown sugar; mix well. Stir in caraway seed and paprika. Simmer for 20 minutes, adding reserved sauerkraut liquid if necessary to make of desired consistency. Yield: 4 servings.

Approx per serving: Cal 448, Prot 19.1 gr,
T Fat 29.8 gr, Chol 70.3 mg, Carbo 27.0 gr,
Sod 2356.6 mg, Potas 530.8 mg.

Juanita Boyce
Lawton H.S., Lawton, OK

Baked Chicken-Rice Casserole

1 c. rice
1 c. whole kernel corn
1 c. green peas
1 can cream of mushroom soup
1 can cream of chicken soup
2 soup cans water
1 tsp. salt
8 pieces chicken
½ c. melted butter

Layer rice, corn and peas in buttered 9x13-inch baking dish. Combine soups, water and salt in bowl; mix well. Pour over layers. Dip chicken into melted butter. Sprinkle with salt to taste. Arrange over layers. Bake at 275 degrees for 2½ hours. Yield: 4 servings.

Approx per serving: Cal 762, Prot 45.3 gr, T Fat 37.0 gr, Chol 172.5 mg, Carbo 61.4 gr, Sod 2227.2 mg, Potas 719.9 mg.

Emily Lewis
Capitol Hill H.S., Oklahoma City, OK

Chicken and Rice Casserole

2½ c. uncooked minute rice
1 chicken, cut up
1 can cream of celery soup
1 can cream of mushroom soup
1 c. milk
1 env. dry onion soup mix

Layer rice and chicken in 9x13-inch baking dish. Spoon mixture of next 3 ingredients over chicken. Sprinkle with soup mix. Cover tightly with foil. Bake at 375 degrees for 2½ hours. Yield: 6 servings.

Approx per serving: Cal 730, Prot 90.6 gr, T Fat 17.3 gr, Chol 232.3 mg, Carbo 46.4 gr, Sod 1458.7 mg, Potas 1271.1 mg.

Mary Anne Fleischacker
Columbus Sr. H.S., Columbus, NE

Chicken with Vegetables

1   2½-lb. chicken, cut up
1 chicken bouillon cube
¾ c. boiling water
1 c. 1-inch carrot chunks
¼ c. chopped onion
1 lb. mushrooms, sliced
½ lb. green beans, cut into 1-in. pieces
1 tsp. bay leaves
⅛ tsp. pepper
2 tbsp. flour
2 tbsp. water

Place chicken on rack in 9x13-inch baking pan. Bake at 450 degrees for 20 minutes or until brown. Remove chicken; drain. Stir bouillon dissolved in boiling water into baking pan. Add vegetables and seasonings. Add chicken. Spoon sauce over chicken. Bake, covered, at 350 degrees until chicken is tender. Arrange chicken and vegetables on serving plate. Blend pan drippings and mixture of flour and 2 tablespoons water in saucepan. Cook until thickened, stirring constantly. Spoon over chicken. Serve with rice. Yield: 4 servings.

Approx per serving: Cal 272, Prot 41.1 gr, T Fat 4.5 gr, Chol 90.3 mg, Carbo 16.3 gr, Sod 346.8 mg, Potas 1217.6 mg.

Patsy McCowan
Eureka, CA

Chicken Casserole

1   3½ to 4 lb. chicken, cut up
½ c. flour
2 tsp. each salt and pepper
½ c. butter
1 c. water
1   8-oz. can tomato sauce
1   4-oz. can mushrooms with liquid
2 tbsp. minced onion
1 tbsp. celery flakes
1 tbsp. green pepper flakes
¼ tsp. nutmeg

Coat chicken with mixture of flour, salt and pepper. Brown on both sides in butter in skillet. Remove chicken to baking dish. Add remaining ingredients to skillet, stirring to deglaze. Simmer for 5 minutes. Pour over chicken. Bake, covered, at 350 degrees for 1 hour. Serve on bed of rice with broccoli spears and whole baby carrots. Yield: 8 servings.

Approx per serving: Cal 262, Prot 23.9 gr, T Fat 14.1 gr, Chol 91.5 mg, Carbo 9.0 gr, Sod 882.8 mg, Potas 421.8 mg.

Beverley C. Goodman
Smyth Co. Voc. Sch., Marion, VA

Country Captain

1¼ lb. chicken pieces
½ tsp. garlic salt
½ tsp. paprika
8 oz. onions, chopped
2 med. green peppers, chopped
¾ tsp. garlic powder
2 c. canned tomatoes
½ c. tomato paste
1 tbsp. parsley flakes
Artificial sweetener to equal 2 tsp. sugar
1 tsp. thyme
1 tsp. curry powder
1 tsp. Sherry extract
¼ tsp. pepper
1 c. cooked rice

Season chicken with garlic salt and paprika. Brown in nonstick skillet over medium heat. Remove chicken. Add onions, green peppers and garlic powder. Stir-fry for 4 minutes. Add tomatoes, tomato paste, parsley and seasonings; mix well. Add chicken. Cook, covered, for 40 minutes or until chicken is tender. Serve over hot rice. Yield: 2 servings.

*Approx per serving: Cal 433, Prot 33.8 gr,
T Fat 4.0 gr, Chol 58.3 mg, Carbo 68.6 gr,
Sod 1312.0 mg, Potas 2139.7 mg.*

*Haley Haas
Spokane, WA*

Crunchy Deviled Chicken

1 tsp. dry mustard
1 tbsp. water
1 egg, beaten
1½ tsp. seasoned salt
1 tsp. parsley flakes
Pepper to taste
1 fryer, cut up
1 to 1½ c. potato flakes
3 tbsp. butter, melted

Blend mustard with water in bowl. Let stand for 10 minutes. Add egg, seasoned salt, parsley flakes and pepper; mix lightly. Dip chicken into egg mixture. Roll in potato flakes, coating generously. Arrange skin side down in melted butter in baking dish. Bake at 400 degrees for 40 to 50 minutes or until tender, turning once. Yield: 4 servings.

*Approx per serving: Cal 278, Prot 24.0 gr,
T Fat 14.1 gr, Chol 167.0 mg, Carbo 14.2 gr,
Sod 993.2 mg, Potas 530.3 mg.*

*Joyce Ritter
Camden, NJ*

Gayla's Baked Chicken

2 chickens, cut up
1 tsp. ginger
⅓ tsp. garlic powder
2 tsp. salt
½ tsp. pepper

Arrange chicken skin side up in single layer in 9x13-inch baking dish. Combine remaining ingredients in shaker; mix well. Sprinkle over chicken. Bake at 325 degrees for 1 hour or until tender. Yield: 8 servings.

*Approx per serving: Cal 235, Prot 44.8 gr,
T Fat 4.8 gr, Chol 112.0 mg, Carbo 0.0 gr,
Sod 623.6 mg, Potas 582.6 mg.*

*Marcia Ingram
Lexington H.S., Lexington, TX*

Golden Chicken

2 tbsp. melted butter
2 lb. chicken pieces
1 can cream of chicken soup
½ c. sliced almonds

Spread butter evenly in 9x13-inch baking dish. Arrange chicken in prepared dish. Bake at 400 degrees for 40 minutes, turning chicken after 20 minutes and 40 minutes. Spread soup over chicken. Sprinkle with almonds. Bake for 20 minutes or until chicken is tender. Serve over creamed potatoes or rice. Yield: 4 servings.

*Approx per serving: Cal 431, Prot 49.3 gr,
T Fat 21.9 gr, Chol 135.7 mg, Carbo 7.8 gr,
Sod 764.1 mg, Potas 744.5 mg.*

*Gail Dixon
Lyons Jr. H.S., Vidalia, GA*

Island-Style Chicken

2 lb. chicken pieces
2 tbsp. oil
1   8-oz. can pineapple chunks
1 can chicken broth
½ c. vinegar
1 tsp. soy sauce
1 lg. clove of garlic, minced
1 med. green pepper, chopped
3 tbsp. cornstarch
¼ c. water

Brown chicken in oil in skillet; drain. Drain pineapple, reserving juice. Combine chicken, reserved juice, broth, vinegar, soy sauce and garlic in skillet. Cook, covered, over low heat for 40 minutes. Add pineapple and green pepper. Cook for 5 minutes, stirring occasionally. Blend cornstarch with water. Stir into skillet. Cook until thickened, stirring constantly. Serve with rice. Yield: 4 servings.

Approx per serving: Cal 231, Prot 17.1 gr, T Fat 10.3 gr, Chol 59.2 mg, Carbo 21.1 gr, Sod 516.3 mg, Potas 323.2 mg.

Celeste Jacob
New Orleans, LA

Juicy Chicken

3 c. finely crushed Cheerios
1 tsp. salt
⅛ tsp. each sage, garlic powder
½ tsp. each pepper, onion powder
½ tsp. each oregano, parsley flakes
1   2½ to 3-lb. chicken, cut up
½ c. milk
2 tbsp. melted butter

Combine Cheerios with seasonings in bowl; mix well. Dip chicken in milk and coat with cereal mixture. Place skin side up in foil-lined baking pan. Drizzle with butter. Bake at 350 degrees for 1 hour or until tender. Do not turn. Yield: 6 servings.

Approx per serving: Cal 177, Prot 16.2 gr, T Fat 7.5 gr, Chol 65.8 mg, Carbo 10.4 gr, Sod 609.3 mg, Potas 191.5 mg.

Karan Heffelfinger
Carbon, PA

Low-Cal Chicken

2½ lb. chicken pieces, skinned
½ c. water
2 pkg. instant chicken broth
    and seasoning mix
½ tsp. mustard
2 tsp. Worcestershire sauce
2 or 3 dashes of hot sauce
1 clove of garlic, minced
2 tsp. curry powder
1 tsp. each oregano, salt
½ tsp. paprika

Arrange chicken pieces in shallow baking dish. Bring water to a boil in saucepan. Add broth mix; stir until mix is dissolved. Add remainnig ingredients; mix well. Pour over chicken. Bake at 350 degrees for 1 hour or until chicken is tender. Yield: 4 servings.

Approx per serving: Cal 194, Prot 36.4 gr, T Fat 3.9 gr, Chol 91.1 mg, Carbo 0.8 gr, Sod 1120.1 mg, Potas 493.2 mg.

Debbie Hampton
Charleston Sr. H.S., Charleston, MO

Almond Chicken

4 chicken breast filets, thinly sliced
2 tbsp. oil
1 sm. onion, thinly sliced
1 c. each sliced celery, water chestnuts
1   5-oz. can bamboo shoots
2 c. chicken broth
2 tbsp. soy sauce
2 tbsp. cornstarch
1 tsp. sugar
¼ c. cold water
¼ c. slivered almonds, toasted

Stir-fry chicken in hot oil in heavy skillet for 5 minutes. Add onion and celery. Stir-fry for 5 minutes. Add water chestnuts, bamboo shoots, chicken broth and soy sauce; mix well. Cook, covered, for 5 minutes. Blend cornstarch, sugar and cold water. Stir into chicken. Cook until thickened, stirring constantly. Sprinkle with almonds. Serve over rice. Yield: 4 servings.

Approx per serving: Cal 353, Prot 31.1 gr, T Fat 16.0 gr, Chol 65.2 mg, Carbo 22.1 gr, Sod 1240.3 mg, Potas 924.0 mg.

Gervis B. Spear
Darlington, SC

Broiled Lemon Chicken

12 wooden skewers
6 lb. chicken breasts
¼ c. oil
½ c. soy sauce
Juice of 2 lemons
½ tsp. white pepper
2 tsp. salt

Place skewers in water to soak. Skin and bone chicken; cut into cubes. Combine with remaining ingredients in bowl. Chill for 2 hours or longer. Drain, reserving marinade. Thread chicken tightly onto skewers. Place on rack in broiler pan or over hot coals. Brush with reserved marinade. Broil for 6 minutes, or until cooked through, turning occasionally. Serve with rice. Yield: 12 servings.

Approx per serving: Cal 341, Prot 56.0 gr,
T Fat 10.6 gr, Chol 138.2 mg, Carbo 1.9 gr,
Sod 1346.4 mg, Potas 777.5 mg.

Jane Bigler
Hillcrest Jr. H.S., Lenexa, KS

Chicken with Cashews

¼ c. peanut oil
4 chicken breast filets, chopped
½ c. chopped green pepper
½ c. cashews
2 tbsp. chopped green onion
2 cloves of garlic, finely chopped
¼ tsp. ginger
¼ c. soy sauce
2 tbsp. dark corn syrup
1 tbsp. vinegar
4 tsp. cornstarch
½ c. water

Heat oil in preheated wok for 2 minutes. Add chicken. Stir-fry for 2 to 3 minutes; push to side. Add green pepper and cashews. Stir-fry for 30 seconds; push to side. Add green onion, garlic and ginger. Stir-fry for 1 minute; push to side. Stir in mixture of remaining ingredients. Cook for 1 minute, stirring constantly. Yield: 6 servings.

Approx per serving: Cal 288, Prot 15.4 gr,
T Fat 19.5 gr, Chol 27.6 mg, Carbo 14.5 gr,
Sod 904.9 mg, Potas 270.1 mg.

Marcia Fehl
York, PA

Chicken Divan

1 10-oz. package frozen broccoli spears
4 chicken breasts, cooked, sliced
1 can Cheddar cheese soup
⅓ c. milk
¼ tsp. each garlic powder, basil
* and celery salt*

Cook broccoli in a small amount of water in saucepan until tender-crisp; drain. Arrange in baking dish. Place chicken slices over broccoli. Combine remaining ingredients in saucepan; mix well. Heat until bubbly. Spoon over chicken. Bake at 375 degrees for 20 minutes or until heated through. Yield: 4 servings.
Note: Nutritional information does not include cheese soup.

Approx per serving: Cal 208, Prot 36.2 gr,
T Fat 4.5 gr, Chol 85.8 mg, Carbo 4.7 gr,
Sod 222.6 mg, Potas 631.7 mg.

Deborah Lanuti
Valley View H.S., Scranton, PA

No-Fuss Baked Chicken

½ c. salad dressing
¼ c. Parmesan cheese
1 tbsp. prepared mustard
1 tbsp. Dijon mustard
4 chicken breast filets
⅔ c. crushed cornflakes

Combine first 4 ingredients in bowl; mix well. Reserve ⅓ cup for use as dipping sauce. Rinse chicken and pat dry. Brush both sides with sauce. Place on lightly greased baking sheet. Sprinkle with cornflake crumbs. Bake at 400 degrees for 20 minutes. Serve with reserved sauce. Yield: 2 servings.
Note: May use reduced-calorie salad dressing and substitute bran flakes for cornflakes.

Approx per serving: Cal 789, Prot 75.3 gr,
T Fat 37.1 gr, Chol 210.0 mg, Carbo 34.3 gr,
Sod 1068.7 mg, Potas 944.3 mg.

Cynthia SuAnn Lutz
Camden-Frontier Sch., Reading, MI

Chicken-Raisin Sauté

8 chicken breast filets
½ tsp. each curry powder, cardamom
1 clove of garlic, chopped
1 shallot, chopped
1 tbsp. oil
½ c. dry vermouth
½ c. chicken stock
¾ c. raisins
½ tsp. cinnamon
½ c. silvered almonds
½ c. chopped parsley

Sprinkle chicken with curry powder and cardamom. Sauté garlic and shallot in oil in skillet. Add chicken. Sauté until lightly browned. Add vermouth and chicken stock. Simmer for several minutes. Stir in raisins and cinnamon. Simmer until raisins are plump. Sprinkle with almonds and parsley. Serve with rice. Yield: 4 servings.

Approx per serving: Cal 588, Prot 70.6 gr,
T Fat 20.2 gr, Chol 166.1 mg, Carbo 26.1 gr,
Sod 204.6 mg, Potas 1251.4 mg.

Rosette Backas
Hinsdale South H.S., Clarendon Hills, IL

Chinese-Style Chicken

4 chicken breast filets, cut into thin strips
2 tbsp. oil
¾ c. each silvered celery, green pepper,
and carrot
½ c. chicken broth
¼ c. soy sauce
1 sm. clove of garlic, chopped
¼ tsp. ginger
2 tbsp. cornstarch
2 tbsp. parsley flakes
Juice of ½ lemon
½ tsp. salt

Stir-fry chicken in oil in wok over high heat for 3 minutes. Add vegetables. Stir-fry for 3 minutes longer. Mix remaining ingredients in small bowl. Pour over chicken and vegetables. Bring to a boil. Cook for 2 minutes longer, stirring constantly. Serve over rice. Yield: 6 servings.

Approx per serving: Cal 382, Prot 31.4 gr,
T Fat 9.0 gr, Chol 74.0 mg, Carbo 41.5 gr,
Sod 1208.1 mg, Potas 457.3 mg.

Annette Wilson
Omaha, NE

Microwave California Chicken

4 chicken breast filets
2 tsp. lemon juice
1 tsp. dried onion flakes
Basil and pepper to taste
⅔ c. shredded sharp Cheddar cheese
½ avocado, thinly sliced
4 thin tomato slices

Arrange chicken breasts in glass baking dish. Sprinkle with lemon juice, onion flakes, basil and pepper. Microwave, covered with waxed paper, on High for 7½ minutes or until chicken is tender, turning dish twice. Sprinkle with half the cheese. Arrange avocado and tomato on top. Microwave, covered, for 2½ minutes. Sprinkle with remaining cheese. Let stand, covered, for 5 minutes. Yield: 4 servings.

Approx per serving: Cal 338, Prot 40.5 gr,
T Fat 15.9 gr, Chol 101.7 mg, Carbo 8.0 gr,
Sod 188.1 mg, Potas 803.3 mg.

Dolly Stewart
Pasadena, CA

Special Chicken

1 c. rice
1 stick margarine
2 cans chicken broth
12 chicken breast filets
4 1-oz. thin slices ham
4 ½-oz. slices Swiss cheese

Brown rice in margarine in heavy skillet. Stir in chicken broth. Pour into 2-quart casserole. Place chicken in rice mixture. Bake at 350 degrees for 1 hour. Layer ham slices and cheese slices over chicken. Bake until cheese melts. Yield: 8 servings.

Approx per serving: Cal 499, Prot 56.9 gr,
T Fat 20.1 gr, Chol 143.6 mg, Carbo 18.8 gr,
Sod 599.9 mg, Potas 718.7 mg.

Wilda Cornett
Heritage H.S., Rockford, TN

Chicken and Vegetable Kabobs

6 chicken breast filets
1 16-oz. bottle of French dressing
1 zucchini
1 green pepper

3 onions
8 lg. mushrooms
2 carrots
1 c. Italian dressing

Cut chicken into 1 x 1½-inch strips. Combine with French dressing to cover in plastic bag; squeeze out air and tie securely. Cut vegetables into chunks. Combine with remaining French dressing in plastic bag; squeeze out air and tie securely. Marinate in refrigerator for several hours to overnight. Thread chicken and vegetables alternately onto skewers. Grill over medium heat until tender, turning and basting frequently with Italian dressing. Serve with chicken-flavored rice. Yield: 4 servings.
Note: Nutritional information includes full amount of marinade and Italian dressing.

Approx per serving: Cal 1201, Prot 54.4 gr,
T Fat 91.5 gr, Chol 124.4 mg, Carbo 45.1 gr,
Sod 3145.0 mg, Potas 1323.6 mg.

Delinda McCormick
Caldwell Co. H.S., Cadiz, KY

Chicken and Dressing

1   8-oz. package herb-seasoned
      stuffing mix
1 stick butter, melted
4 chicken breasts, cooked
1 can cream of mushroom soup
1 can cream of celery soup
2 cans chicken broth

Toss stuffing mix and butter in bowl. Spread ¾ inch mixture in 9 x 13-inch baking dish. Place chicken over stuffing. Spoon mixture of soups and broth over chicken. Top with remaining stuffing. Bake at 350 degrees for 40 minutes or until bubbly. Yield: 4 servings.

Approx per serving: Cal 728, Prot 43.7 gr,
T Fat 37.9 gr, Chol 168.9 mg, Carbo 53.1 gr,
Sod 3008.1 mg, Potas 666.8 mg.

Ina Pack
Christiansburg H.S., Radford, VA

Chicken Piccata

1 egg
1 tbsp. lemon juice
¼ c. flour
⅛ tsp. each garlic powder, paprika

4 chicken breast filets
¼ c. butter
2 chicken bouillon cubes
½ c. boiling water
2 tbsp. lemon juice

Beat egg with 1 tablespoon lemon juice in small bowl. Combine flour, garlic powder and paprika in bowl; mix well. Dip chicken pieces in egg mixture; coat with flour mixture. Sauté chicken in butter in skillet until brown. Dissolve bouillon cubes in boiling water. Add to chicken with 2 tablespoons lemon juice. Simmer, covered, for 20 minutes or until tender. Yield: 4 servings.

Approx per serving: Cal 345, Prot 36.8 gr,
T Fat 18.0 gr, Chol 183.2 mg, Carbo 7.1 gr,
Sod 686.1 mg, Potas 366.2 mg.

Kim Leissler
Plano, TX

Marvelous Chicken Pie

1 broiler-fryer, cut up
1 tsp. salt
½ c. shredded mozzarella cheese
1   6-oz. can tomato paste
1 tsp. oregano
½ tsp. basil
½ c. small curd cottage cheese
⅔ c. buttermilk baking mix
1 c. milk
2 eggs
¼ tsp. pepper
½ c. shredded mozzarella cheese

Simmer chicken with salt in water to cover in covered saucepan for 45 minutes or until tender; drain. Chop chicken, discarding skin and bones. Mix with ½ cup mozzarella, tomato paste, oregano and basil in bowl. Spread cottage cheese in lightly greased deep-dish pie plate. Spoon chicken mixture over cottage cheese. Beat baking mix, milk, eggs and pepper in bowl for 1 minute. Pour over chicken. Bake at 350 degrees for ½ hour. Sprinkle with remaining ½ cup cheese. Let stand for 5 minutes before cutting. Yield: 6 servings.

Approx per serving: Cal 295, Prot 26.4 gr,
T Fat 12.8 gr, Chol 161.6 mg, Carbo 17.7 gr,
Sod 744.3 mg, Potas 535.0 mg.

Lyndra R. Kolodziej
Floresville, TX

Chicken-Broccoli Casserole

1 chicken, cooked, boned
2   10-oz. packages frozen chopped
*   broccoli, thawed, drained*
1   8-oz. can sliced water chestnuts
1 can mushroom soup
½ c. mayonnaise
1 c. grated Swiss cheese
1   3-oz. can French-fried onions

Layer chicken, broccoli and water chestnuts in baking dish. Combine soup and mayonnaise in bowl; mix well. Spread over layers. Top with cheese. Bake at 350 degrees for 35 minutes. Sprinkle with onions. Bake for 10 minutes longer. Yield: 8 servings.

Approx per serving: Cal 392, Prot 27.0 gr,
T Fat 22.7 gr, Chol 75.3 mg, Carbo 20.4 gr,
Sod 697.5 mg, Potas 512.5 mg.

Marla Kurtz
Knightstown H.S., Knightstown, IN

Chicken Gumbo

1   5-lb. chicken
7 qt. water
1 onion
2 stalks celery
3 chicken bouillon cubes
2 cloves of garlic, chopped
3 c. chopped onion
½ bunch green onions, chopped
2 green peppers, chopped
1 bunch celery, chopped
¾ stick margarine
½ c. flour
2½ c. canned tomatoes
1   2-lb. package frozen cut okra
1 c. rice
½ c. chopped parsley
2 to 3 c. mixed vegetable juice cocktail

Combine chicken with 7 quarts water, 1 onion, 2 stalks celery and bouillon cubes in 8-quart stockpot. Simmer until chicken is tender. Strain 6 quarts stock and reserve. Cool chicken. Bone and chop chicken; measure 4 cups. Sauté garlic, chopped onion, green onions, chopped green peppers and chopped celery in margarine in skillet. Add flour. Cook for 3 minutes, stirring constantly. Combine with reserved stock, tomatoes and salt and pepper to taste in stockpot.

Simmer for 45 minutes. Add okra, rice, chicken, parsley and vegetable juice cocktail. Simmer for 20 to 30 minutes or to desired consistency. Ladle into soup bowls. Yield: 16 servings.

Approx per serving: Cal 241, Prot 15.4 gr,
T Fat 8.0 gr, Chol 31.0 mg, Carbo 28.3 gr,
Sod 467.3 mg, Potas 737.2 mg.

Patricia Mikulecky
Bartlesville Mid.-H.S., Bartlesville, OK

Hot Chicken Salad

4 c. chopped cooked chicken
2 c. chopped celery
1 tsp. chopped onion
⅔ c. slivered almonds
1½ c. grated Cheddar cheese
4 hard-boiled eggs, chopped
1 c. mayonnaise
½ c. chicken broth
1½ tbsp. chopped pimento
1½ tsp. MSG
½ c. bread crumbs
¼ tsp. paprika

Combine first 10 ingredients in bowl; mix well. Spoon into baking dish. Sprinkle bread crumbs and paprika over top. Bake at 350 degrees for 30 minutes. Yield: 8 servings.

Approx per serving: Cal 520, Prot 34.0 gr,
T Fat 39.8 gr, Chol 222.6 mg, Carbo 6.4 gr,
Sod 669.2 mg, Potas 536.3 mg.

Valerie Wall
Wichita Northwest H.S., Wichita, KS

Crispy Chicken Casserole

1   2½ to 3-lb. chicken, cut up
2 c. chopped celery
1 lg. onion, chopped
2 chicken bouillon cubes
¼ tsp. each salt and pepper
1 can cream of chicken soup
1 can cream of mushroom soup
3 tbsp. melted margarine
½   8-oz. package corn bread stuffing mix

Combine chicken with celery, onion, bouillon, salt, pepper and water to cover in saucepan. Simmer until chicken is tender. Strain, reserving chicken and vegetables. Chop cooled chicken.

Combine chicken, vegetables and soups in bowl; mix well. Spoon into 9x13-inch baking dish. Mix margarine and stuffing mix in bowl. Sprinkle over chicken. Bake at 350 degrees for 30 minutes. Yield: 4 servings.
Note: Nutritional information does not include corn bread stuffing mix.

Approx per serving: Cal 482, Prot 49.6 gr, T Fat 23.1 gr, Chol 125.4 mg, Carbo 17.4 gr, Sod 2083.9 mg, Potas 968.6 mg.

Kathleen Burchett
Area Supr. of H.E., State Dept. of Ed., Abingdon, VA

Chicken and Wild Rice Casserole

> 1   8-oz. package long grain
>   and wild rice mix
> 2 c. chopped cooked chicken
> 1   16-oz. package frozen French-style
>   green beans
> 1 can cream of celery soup
> ½ c. mayonnaise
> ½ c. sliced water chestnuts
> 2 tbsp. chopped onion
> 2 tbsp. chopped pimento
> ¼ tsp. salt
> Pinch of pepper
> ½ tsp. paprika

Prepare rice according to package directions. Combine with chicken, beans, soup, mayonnaise, water chestnuts, onion, pimento, salt and pepper in bowl; mix well. Spoon into baking dish. Sprinkle with paprika. Bake at 350 degrees for 25 to 30 minutes or until bubbly. Yield: 12 servings.
Note: Nutritional information does not include rice mix.

Approx per serving: Cal 144, Prot 8.7 gr, T Fat 9.4 gr, Chol 26.5 mg, Carbo 6.7 gr, Sod 315.3 mg, Potas 201.7 mg.

Sue Smith
Bartlesville H.S., Bartlesville, OK

Delicious Chicken

> 2   10-oz. packages frozen broccoli,
>   cooked
> 4 chicken breasts, cooked, sliced
> 2 cans cream of chicken soup
> 1 tsp. lemon juice

> 1 tsp. curry powder
> ¾ c. reduced-calorie salad dressing
> 4 oz. sliced mushrooms
> 1 c. seasoned stuffing mix
> ¾ c. shredded sharp Cheddar cheese

Arrange broccoli in 9x13-inch baking pan. Top with chicken. Combine next 5 ingredients in bowl; mix well. Pour over chicken. Sprinkle dry stuffing mix and cheese over top. Bake at 350 degrees for 20 to 30 minutes. Yield: 6 servings.

Approx per serving: Cal 325, Prot 32.8 gr, T Fat 12.5 gr, Chol 77.7 mg, Carbo 20.8 gr, Sod 1118.7 mg, Potas 614.2 mg.

Carol Kizer
DeRidder, LA

Baked Turkey Thermidor

> 1   10-oz. package frozen green peas,
>   cooked, drained
> 2 c. chopped cooked turkey
> 1 c. chopped celery
> 1   5-oz. can sliced water chestnuts,
>   drained
> ½ c. sliced toasted almonds
> 2 tbsp. chopped green pepper
> 1 tbsp. grated onion
> 2 tbsp. chopped pimento
> 2 tbsp. white wine
> 1 tbsp. lemon juice
> ½ tsp. salt
> 1 can cream of chicken soup
> ½ c. milk
> 2 slices bread, cubed
> 1 c. grated Cheddar cheese

Combine first 8 ingredients in 2-quart casserole. Sprinkle with wine, lemon juice and salt. Blend soup and milk in saucepan. Bring to a boil over low heat, stirring constantly. Add to turkey mixture; mix well. Sprinkle bread cubes over top. Bake at 350 degrees until bread is lightly browned. Top with cheese. Bake until cheese is melted. Yield: 12 servings.

Approx per serving: Cal 211, Prot 18.8 gr, T Fat 8.9 gr, Chol 41.8 mg, Carbo 13.4 gr, Sod 461.5 mg, Potas 343.5 mg.

Barbara Huff
Columbus, GA

Turkey Casserole

1   7-oz. package elbow macaroni
1½ lb. ground fresh turkey
1 onion, chopped
1   32-oz. jar salt-free spaghetti sauce
2 c. grated mozzarella cheese

Cook macaroni according to package directions; drain. Brown turkey and onion in skillet. Stir in macaroni and spaghetti sauce; mix well. Spoon into 9 x 13-inch baking dish. Sprinkle with cheese. Bake at 350 degrees for 30 minutes. Yield: 8 servings.
Note: Nutritional information does not include spaghetti sauce.

Approx per serving: Cal 310, Prot 21.9 gr, T Fat 14.7 gr, Chol 73.2 mg, Carbo 21.3 gr, Sod 159.2 mg, Potas 255.1 mg.

Marian E. Baker
Sycamore H.S., Sycamore, IL

Turkey Gumbo

1 lb. ground beef
2 lg. onions, chopped
3 stalks celery, chopped
2   28-oz. cans tomatoes, chopped
2   10-oz. packages frozen okra
1   6-oz. can shrimp
3 c. chicken bouillon
7 c. chopped cooked turkey
3 tbsp. gumbo filé
½ tsp. salt
Pepper to taste

Cook ground beef, onions and celery in large stockpot for 25 minutes, stirring frequently; drain. Add tomatoes, okra, shrimp, bouillon and turkey. Stir in filé, salt and pepper. Simmer for 1 to 1½ hours. Serve over rice. Yield: 12 servings.

Approx per serving: Cal 416, Prot 12.6 gr, T Fat 6.0 gr, Chol 47.2 mg, Carbo 81.0 gr, Sod 433.6 mg, Potas 509.8 mg.

Beckie Mischke
Los Alamos County, NM

Turkey Roll

1½ lb. ground fresh turkey
½ c. finely chopped onion
¼ c. tomato sauce
2 slices bread, crumbled
1 tsp. dry mustard
½ tsp. oregano
2 eggs, beaten
¼ tsp. garlic powder
1   10-oz. package frozen spinach, thawed
½ c. shredded mozzarella cheese
¾ c. tomato sauce

Combine first 8 ingredients in bowl; mix well. Pat into 8x12-inch rectangle on foil. Spread spinach over turkey mixture. Sprinkle with cheese. Roll from narrow side as for jelly roll. Place seam side down in ungreased baking pan. Bake at 350 degrees for 1 hour. Let stand for 10 minutes. Heat ¾ cup tomato sauce in saucepan. Place turkey roll on serving platter. Serve with warm tomato sauce. Yield: 6 servings.

Approx per serving: Cal 194, Prot 15.6 gr, T Fat 9.8 gr, Chol 125.0 mg, Carbo 11.4 gr, Sod 381.1 mg, Potas 483.7 mg.

Crystal Warner
Rugby, ND

Turkey Scallopini

12   ¼-in. thick turkey breast slices
½ c. flour
½ tsp. white pepper
2 tbsp. butter
¾ c. chicken stock
6 paper-thin lemon slices
1 tbsp. lemon juice

Coat turkey with flour and pepper. Sauté 3 or 4 at a time in butter in skillet for 2 minutes on each side. Remove turkey. Add stock. Boil for 1 to 2 minutes, stirring constantly. Arrange turkey in skillet; top with lemon slices. Simmer, covered, for 10 to 15 minutes or until tender. Arrange turkey and lemon slices on warm serving platter. Add lemon juice to pan drippings. Bring to a boil, stirring constantly to deglaze skillet. Pour over turkey. Yield: 6 servings.

Approx per serving: Cal 175, Prot 20.0 gr, T Fat 6.2 gr, Chol 56.0 mg, Carbo 8.9 gr, Sod 253.6 mg, Potas 260.4 mg.

Maureen Munro Dillard
Los Angeles, CA

Crunchy Baked Fish

1 lb. fish fillets
⅓ c. cornflake crumbs
¼ tsp. onion powder
1 tbsp. chopped parsley
½ c. low-fat yogurt

Cut fish into serving portions; rinse and pat dry. Combine cornflake crumbs, onion powder and parsley in bowl; mix well. Dip fish into yogurt. Coat with crumb mixture. Arrange in single layer in lightly oiled shallow baking dish. Bake at 500 degrees for 10 minutes. Garnish with lemon slices. Yield: 4 servings.

Approx per serving: Cal 217, Prot 22.9 gr,
T Fat 9.9 gr, Chol 64.7 mg, Carbo 7.5 gr,
Sod 143.9 mg, Potas 387.8 mg.

Peggy S. Christian
Hampton H.S., Carrollton, VA

Fish Cakes

5 tbsp. flour
½ tsp. salt
5 tbsp. melted butter
1½ c. milk
1½ tbsp. minced onion
1½ tbsp. minced parsley
1 tsp. lemon juice
3 c. flaked cooked fish
1 egg, beaten
2 tbsp. water
1 c. fine dry bread crumbs
2 tbsp. butter

Blend flour and salt with 5 tablespoons butter in saucepan. Add milk gradually, stirring constantly. Cook until thickened, stirring constantly. Stir in onion, parsley, lemon juice and fish. Chill in refrigerator. Shape into 12 cakes. Mix egg with water in bowl. Roll fish cakes in bread crumbs. Dip in egg mixture then into bread crumbs. Chill for 1 hour. Cook in 2 tablespoons butter in skillet until golden brown. Yield: 6 servings.

Approx per serving: Cal 301, Prot 24.9 gr,
T Fat 17.2 gr, Chol 92.2 mg, Carbo 10.8 gr,
Sod 406.6 mg, Potas 125.1 mg.

Eileen Graham
Adams, PA

Flounder Creole

1 c. sliced green onions with tops
1 lg. green pepper, sliced into thin strips
1½ tbsp. margarine
1 16-oz. can tomatoes
1 8-oz. can tomato sauce
1 bay leaf
½ tsp. thyme
6 4-oz. flounder fillets
¼ tsp. salt
⅛ tsp. pepper

Sauté green onions and green pepper in margarine in skillet until tender. Drain and chop tomatoes, reserving liquid. Stir tomatoes, reserved liquid, tomato sauce, bay leaf and thyme into skillet. Simmer for 20 minutes. Discard bay leaf. Arrange fillets in 9x13-inch baking dish; sprinkle with salt and pepper. Spoon sauce over fillets. Bake, covered, at 350 degrees for 20 minutes. Serve over rice. Yield: 6 servings.

Approx per serving: Cal 167, Prot 35.9 gr,
T Fat 4.1 gr, Chol 56.3 mg, Carbo 9.0 gr,
Sod 533.9 mg, Potas 1079.9 mg.

Phyllis Grainger
Reno, NV

Creole Haddock

1 lb. haddock fillets
¼ c. chopped onion
2 tbsp. chopped green pepper
1 tbsp. margarine
1 c. chopped stewed tomatoes
¼ tsp. oregano

Arrange fillets in baking dish. Sauté onion and green pepper in margarine in skillet. Add tomatoes, oregano and salt and pepper to taste. Simmer for several minutes. Spoon over fillets. Bake at 350 degrees for 40 minutes or until fish flakes easily. Yield: 4 servings.

Approx per serving: Cal 141, Prot 18.6 gr,
T Fat 3.1 gr, Chol 58.0 mg, Carbo 3.3 gr,
Sod 184.5 mg, Potas 464.9 mg.

Carolyn Seedman
Long Island, NY

Lemon Haddock

2 16-oz. packages frozen haddock
  fillets, thawed
½ tsp. salt
1 c. sliced fresh mushrooms
¼ c. chopped onion
¼ c. chopped green pepper
8 thin lemon slices
¼ c. dry Sauterne
Paprika to taste

Arrange fillets in greased baking dish; sprinkle with salt. Layer mushrooms, onion, green pepper and lemon slices over fillets. Drizzle wine over top. Sprinkle with paprika. Bake, covered, at 350 degrees for 30 minutes or until fish flakes easily. Yield: 8 servings.

Approx per serving: Cal 110, Prot 18.3 gr, T Fat 0.2 gr, Chol 57.9 mg, Carbo 1.8 gr, Sod 206.0 mg, Potas 382.3 mg.

Sherrie Fink
Allentown, PA

Baked Halibut

1 onion, thinly sliced
2 lb. halibut fillets
1½ c. sliced fresh mushrooms
⅓ c. chopped tomato
¼ c. chopped green pepper
¼ c. chopped parsley
½ c. dry white wine
2 tbsp. lemon juice
¼ tsp. dillweed
⅛ tsp. pepper

Layer onion, fillets, mushrooms, tomato, green pepper and parsley in baking dish. Drizzle mixture of wine, lemon juice and seasonings over layers. Bake at 350 degrees for 30 minutes or until fish flakes easily. Yield: 6 servings.

Approx per serving: Cal 179, Prot 31.9 gr, T Fat 1.9 gr, Chol 73.0 mg, Carbo 5.2 gr, Sod 90.8 mg, Potas 825.8 mg.

Shannon Poteet
St. Louis, MO

Poached Halibut

½ med. onion, sliced
1 tbsp. lemon juice
Chopped parsley to taste
1 bay leaf
1 tsp. salt
3 to 6 peppercorns
2 to 3 c. water
1 lb. halibut fillets

Combine first 6 ingredients in saucepan. Add 2 to 3 cups water. Simmer for 5 minutes. Arrange fillets in saucepan. Simmer for 5 minutes or until fish flakes easily. Do not overcook. Remove fish to serving plate. Garnish with lemon wedges, tomato slices and fresh parsley. Yield: 4 servings.

Approx per serving: Cal 115, Prot 23.0 gr, T Fat 1.4 gr, Chol 54.8 mg, Carbo 2.2 gr, Sod 596.9 mg, Potas 517.8 mg.

Candice Thompson
Lombard, IL

Salmon Pâté

1 16-oz. can pink salmon
2 env. unflavored gelatin
½ c. reduced-calorie mayonnaise
½ c. chili sauce
1 tbsp. Worcestershire sauce
2 tbsp. lemon juice
1 tsp. dillweed
1 7-oz. can water-pack tuna, drained
4 hard-boiled eggs, chopped
¼ c. chopped onion
¼ c. chopped stuffed olives

Drain salmon, reserving liquid. Soften gelatin in reserved liquid in saucepan. Heat until gelatin dissolves, stirring constantly. Add next 5 ingredients; mix well. Stir in salmon, tuna, eggs, onion and olives. Pour into fish-shaped mold. Chill until set. Unmold onto lettuce-lined serving plate. Serve with assorted crackers. Yield: 60 tablespoons.

Approx per tablespoon: Cal 30, Prot 3.1 gr, T Fat 1.5 gr, Chol 22.0 mg, Carbo 0.8 gr, Sod 105.7 mg, Potas 51.8 mg.

Elizabeth Eberle
Topeka, KS

Salmon Loaf

½ c. low-fat milk
1½ slices bread, crumbled
1 c. flaked pink salmon
¼ tsp. salt
⅛ tsp. paprika
⅛ tsp. pepper
2 tsp. lemon juice
2 egg whites, stiffly beaten

Mix milk and bread crumbs in double boiler. Heat until warm. Add salmon, seasonings and lemon juice; mix well. Fold gently into egg whites. Pour into casserole; place in larger pan of hot water. Bake at 300 degrees for 1 hour.
Yield: 4 servings.

Approx per serving: Cal 129, Prot 15.0 gr, T Fat 3.9 gr, Chol 20.8 mg, Carbo 7.6 gr, Sod 439.4 mg, Potas 291.0 mg.

> *Mary Bergmann*
> *San Angelo, TX*

Salmon and Zucchini Parmigiana

1 c. sliced mushrooms
¼ c. chopped onion
1 clove of garlic, minced
2 tbsp. oil
3 c. diagonally sliced zucchini
¼ tsp. oregano
¼ tsp. basil
1   7¾-oz. can salmon, drained
1 c. chopped tomato
½ c. shredded mozzarella cheese
3 tbsp. Parmesan cheese
1 tbsp. minced parsley

Sauté mushrooms, onion and garlic in oil in skillet. Add zucchini, oregano and basil. Cook until zucchini is tender-crisp. Add salmon and tomato; mix lightly. Cook for 1 minute longer. Sprinkle cheeses around edge of skillet. Sprinkle parsley in center. Cook, covered, for 2 minutes or until cheese is melted. Yield: 3 to 4 servings.

Photograph for this recipe is on Cover.

Salmon-Stuffed Pasta Shells

12 jumbo macaroni shells
1 c. drained canned salmon
1 egg, beaten

1 c. ricotta cheese
2 tbsp. chopped onion
2 tbsp. minced parsley
¼ tsp. finely grated lemon rind
½ c. light cream
1½ tbsp. butter
1½ tbsp. flour
¼ tsp. salt
⅛ tsp. pepper
1½ c. milk
3 tbsp. minced fresh dill
*   or 2 tsp. dried dill*
1 tbsp. lemon juice

Cook pasta according to package directions; drain. Spread on waxed paper to cool slightly. Combine salmon, egg, ricotta cheese, onion, parsley and lemon rind in bowl; mix well. Spoon into pasta shells. Pour cream into buttered 9-inch square baking dish. Arrange shells in dish. Cover with foil. Bake at 350 degrees for 30 minutes or until bubbly. Melt butter in small saucepan over medium heat. Blend in flour, salt and pepper; remove from heat. Add milk gradually. Cook over medium heat until thickened, stirring constantly. Stir in dill and lemon juice. Arrange shells on serving plate. Serve with dill sauce.
Yield: 4 to 6 servings.

Photograph for this recipe on page 4.

Snappy Snapper

12 oz. red snapper fillets
2 tsp. mayonnaise
1 tsp. Dijon mustard
1 tsp. lemon juice
1 tsp. chopped chives
1 tsp. chopped parsley
2 tsp. Parmesan cheese

Arrange fillets in shallow flameproof baking dish sprayed with nonstick cooking spray. Combine mayonnaise, mustard, lemon juice, chives and parsley in small bowl; mix well. Spread over fillets. Sprinkle with cheese. Bake at 400 degrees for 20 minutes. Broil for 1 minute.
Yield: 2 servings.

Approx per serving: Cal 216, Prot 39.0 gr, T Fat 2.5 gr, Chol 105.5 mg, Carbo 0.5 gr, Sod 196.7 mg, Potas 1105.9 mg.

> *Pat Barnes*
> *Gretna, LA*

Stuffed Trout

1 sm. onion, chopped
2 tbsp. butter
2 c. canned crab meat
½ tsp. salt
Freshly ground pepper
8 trout, rinsed, dried
1 stick butter, melted
Chopped parsley
1 tsp. lemon juice

Sauté onion in 2 tablespoons butter in skillet until tender. Stir in crab meat; remove from heat. Add salt, pepper and enough water to moisten. Spoon into trout. Place in well-buttered baking dish. Top with remaining ingredients. Bake at 450 degrees for 20 minutes. Yield: 8 servings.

Approx per serving: Cal 471, Prot 72.0 gr,
T Fat 17.8 gr, Chol 193.6 mg, Carbo 1.4 gr,
Sod 752.2 mg, Potas 713.1 mg.

Beth Barclay
Adams, TN

Tuna and Broccoli Bake

¼ c. melted butter
¼ c. flour
1 c. milk
½ tsp. salt
⅛ tsp. Tabasco sauce
1 tsp. Worcestershire sauce
2 tsp. grated onion
½ c. cubed cheese
1   10-oz. package frozen broccoli,
    cooked, drained
1   6-oz. can tuna
Paprika to taste

Blend butter and flour in heavy saucepan. Stir in milk gradually. Cook until thickened, stirring constantly. Stir in salt, Tabasco sauce, Worcestershire sauce, onion and cheese. Cook until cheese melts. Layer broccoli and tuna in buttered casserole. Spoon cheese sauce over top. Sprinkle with paprika. Bake at 325 degrees for 30 minutes. Yield: 4 servings.

Approx per serving: Cal 380, Prot 20.0 gr,
T Fat 27.9 gr, Chol 83.3 mg, Carbo 13.4 gr,
Sod 929.3 mg, Potas 432.5 mg.

Winnie Lang
Dallas, TX

Tuna Imperial

2   7½-oz. cans oil-pack chunk tuna
¼ c. each chopped green onion, celery
⅔ c. flour
¼ tsp. grated lemon rind
2 c. light cream
1   4-oz. can sliced mushrooms
1   5-oz. can sliced water chestnuts
¼ tsp. salt
2 tbsp. parsley

Drain tuna, reserving oil. Sauté green onion and celery in reserved oil in saucepan. Stir in flour and lemon rind. Add cream slowly, stirring until well blended. Cook over low heat until thickened, stirring constantly. Drain mushrooms, reserving liquid. Add enough water to reserved liquid to measure 1¼ cups. Stir into cream sauce. Add water chestnuts, mushrooms, tuna and salt; mix well. Simmer until bubbly. Sprinkle with parsley. Serve with rice. Yield: 6 servings.

Approx per serving: Cal 341, Prot 19.3 gr,
T Fat 22.1 gr, Chol 68.1 mg, Carbo 15.4 gr,
Sod 625.4 mg, Potas 384.9 mg.

Annette Wright
Philadelphia, PA

Barbecued Tuna

2   6-oz. cans water-pack tuna, drained
1 c. finely chopped celery
½ c. chopped onion
1 c. Sugar-Free Barbecue Sauce

Combine tuna, celery, onion and enough Barbecue Sauce to make of desired consistency in saucepan. Simmer for 35 to 45 minutes. Serve in whole wheat pita pockets if desired. Yield: 4 servings.

Sugar-Free Barbecue Sauce

1   46-oz. can tomato juice
6 chicken bouillon cubes
1⅛ c. lemon juice
¾ c. Worcestershire sauce
5 cloves of garlic, chopped
2 tbsp. dry mustard
2 tbsp. dried onion flakes
2 tsp. paprika
1 tsp. pepper
Cayenne pepper to taste
Artificial sweetener to equal ½ c. sugar

Combine all ingredients in saucepan; mix well. Simmer, loosely covered, for 2 hours. Cool. Store in airtight container in refrigerator. Yield: 8 cups.

Approx per serving: Cal 195, Prot 29.2 gr, T Fat 1.1 gr, Chol 60.3 mg, Carbo 18.0 gr, Sod 1368.3 mg, Potas 1030.1 mg.

Jean Beckett
Freeport, IL

Tuna Bake

2 tbsp. margarine, melted
3 tbsp. flour
1 c. low-fat milk
4 eggs, separated
1   7-oz. can water-pack tuna, drained
¼ c. shredded Cheddar cheese
1   10-oz. package frozen
**   chopped broccoli, thawed**
3 green onions, chopped
1 tbsp. mustard
¼ tsp. hot pepper sauce

Blend margarine and flour in saucepan. Stir in milk gradually. Cook until thickened, stirring constantly. Cool slightly. Add egg yolks, tuna, cheese, broccoli, green onions, mustard, hot sauce and salt to taste; mix well. Fold into stiffly beaten egg whites. Spoon into 6-cup soufflé dish. Bake at 350 degrees for 45 minutes or until brown and firm. Yield: 4 servings.

Approx per serving: Cal 294, Prot 26.3 gr, T Fat 15.3 gr, Chol 291.2 mg, Carbo 13.0 gr, Sod 291.1 mg, Potas 503.1 mg.

Maxine Barker
Houston, TX

Tuna Oriental

¼ c. chopped onion
½ clove of garlic, minced
1 tbsp. margarine
2 c. shredded Chinese cabbage
1 tomato, peeled, chopped
1   6-oz. can water-pack tuna, drained
2 c. cooked noodles
½ tsp. salt

Stir-fry onion and garlic in margarine in skillet. Add cabbage. Stir-fry for 5 minutes. Add remaining ingredients. Cook for 5 to 10 minutes, stirring frequently. Spoon into serving bowl. Yield: 4 servings.

Approx per serving: Cal 200, Prot 17.2 gr, T Fat 4.6 gr, Chol 53.8 mg, Carbo 22.3 gr, Sod 332.6 mg, Potas 352.9 mg.

Connie Roy
Bloomington, IN

Baked Fish Italiano

8   4-oz. whitefish fillets
Onion and garlic powder to taste
¼ c. finely chopped celery
¼ c. finely chopped green onions
**   with tops**
¼ c. reduced-calorie Italian dressing
¼ c. lemon juice

Sprinkle fillets with salt, pepper, onion and garlic powder to taste. Arrange in single layer in shallow baking dish. Sprinkle with mixture of remaining ingredients. Bake, covered with foil, at 350 degrees for 20 minutes or until fish flakes easily. Yield: 8 servings.

Approx per serving: Cal 183, Prot 21.5 gr, T Fat 13.9 gr, Chol 62.3 mg, Carbo 1.2 gr, Sod 121.9 mg, Potas 370.5 mg.

Regina Boehler
Mililani, HI

Fish in Foil

4   4-oz. whitefish fillets, ½ in. thick
½ c. buttermilk salad dressing
2 c. broccoli flowerets
1 green pepper, cut into strips
1 sm. onion, thinly sliced

Place 1 fillet on each of four 12-inch foil squares. Top with 2 tablespoons salad dressing and ¼ of the mixture of vegetables. Seal foil. Place on baking sheet. Bake at 450 degrees for 20 minutes. Let stand for 1 to 2 minutes. Yield: 4 servings.

Approx per serving: Cal 334, Prot 27.1 gr, T Fat 20.7 gr, Chol 72.2 mg, Carbo 11.6 gr, Sod 189.9 mg, Potas 911.6 mg.

June Westmere
Findlay, OH

Fish with Tomato Topping

2 lb. whitefish fillets
1   8-oz. can tomato sauce
1 tbsp. onion flakes
1 tbsp. green pepper flakes
½ tsp. oregano
¼ tsp. garlic powder
2 c. shredded mozzarella cheese

Arrange fillets in 9x13-inch baking dish. Combine tomato sauce and seasonings in small bowl; mix well. Spread over fish. Bake at 350 degrees for 35 minutes. Sprinkle with cheese. Bake for 5 minutes or until cheese is melted. Yield: 6 servings.

Approx per serving: Cal 376, Prot 37.6 gr, T Fat 34.1 gr, Chol 116.4 mg, Carbo 6.2 gr, Sod 244.3 mg, Potas 731.7 mg.

Sue Lawson
Haworth Sch., Idabel, OK

Tomato-Clam Bake

1   7-oz. can minced clams
2 c. bread crumbs
¼ c. chopped parsley
1   20-oz. can tomatoes, chopped
¼ c. oil

Drain clams, reserving juice. Alternate layers of clams and bread crumbs in greased 1½-quart baking dish. Sprinkle with parsley and salt and pepper to taste. Spread tomatoes over top. Mix reserved clam juice with oil in small bowl. Pour over layers. Bake at 250 degrees for 20 minutes. Yield: 4 servings.

Approx per serving: Cal 376, Prot 12.2 gr, T Fat 16.6 gr, Chol 20.1 mg, Carbo 44.7 gr, Sod 867.5 mg, Potas 487.9 mg.

Auburn Roswell
Provo, UT

Crab Imperial

1 tbsp. melted margarine
1 tbsp. flour
½ c. milk
1 tsp. minced onion
1½ tsp. Worcestershire sauce
2 slices bread, cubed
½ c. mayonnaise

1 tbsp. lemon juice
½ tsp. salt
Dash of pepper
2 tbsp. margarine
1 lb. crab meat
Paprika to taste

Blend 1 tablespoon margarine and flour in saucepan. Stir in milk gradually. Cook over medium heat until thickened, stirring constantly. Add onion, Worcestershire sauce and bread; mix well. Cool. Blend in mayonnaise, lemon juice, salt and pepper. Brown 2 tablespoons margarine in saucepan. Add crab meat; toss lightly. Stir into sauce. Spoon into greased 1-quart casserole. Sprinkle with paprika. Bake at 350 degrees for 15 minutes or until light brown. Yield: 6 servings.

Approx per serving: Cal 300, Prot 15.1 gr, T Fat 23.1 gr, Chol 91.8 mg, Carbo 8.0 gr, Sod 588.3 mg, Potas 198.9 mg.

Molly Pendleton
Topeka, KS

Crab Stuffing for Flounder

1   4-oz. can mushroom pieces, drained
3 tbsp. chopped onion
1   7-oz. can crab meat, drained
¼ c. dry bread crumbs
1 tbsp. parsley flakes
2 lb. flounder fillets
3 tbsp. flour
1½ c. low-fat milk
¼ c. dry white wine
½ c. shredded Swiss cheese
Paprika to taste

Combine first 5 ingredients and salt and pepper to taste in bowl; mix well. Cut fillets into 8 pieces. Spread each with mushroom mixture; roll to enclose filling. Place seam side down in shallow baking dish. Combine flour, milk and wine in saucepan. Cook over medium heat until thickened, stirring constantly. Add cheese. Cook until cheese melts, stirring constantly. Spoon over rolls. Sprinkle with paprika. Bake at 400 degrees for 30 minutes or until fish flakes easily. Yield: 8 servings.

Approx per serving: Cal 174, Prot 38.4 gr, T Fat 3.6 gr, Chol 62.4 mg, Carbo 7.6 gr, Sod 185.7 mg, Potas 759.7 mg.

Jane Laramie
Clarksburg, WV

Hawaiian Hot Crab Salad

¾ c. pineapple chunks
¾ c. sliced celery
1 tsp. minced onion
1½ tsp. lemon juice
⅓ c. mayonnaise
3 tbsp. chopped almonds
1 tbsp. chopped pimento
1 c. crab meat

Combine first 7 ingredients in bowl. Fold in crab meat. Spoon into casserole. Bake at 350 degrees for 30 minutes or until heated through. Yield: 4 servings.

Approx per serving: Cal 249, Prot 7.8 gr,
T Fat 15.4 gr, Chol 47.4 mg, Carbo 26.7 gr,
Sod 478.9 mg, Potas 226.0 mg.

Bridget Kittson
Marshall, MN

Oysters Rockefeller

Rock salt
36 fresh oysters on the half shell
¼ c. cream
4 med. onions, chopped
2 stalks celery, chopped
8 oz. spinach, chopped
3 sprigs of parsley
1 tsp. salt
¼ tsp. pepper
¼ tsp. cayenne pepper

Spread rock salt in 2 shallow baking dishes. Arrange oysters in shells on rock salt. Place remaining ingredients in blender container. Process until puréed. Spoon purée onto oysters. Bake at 450 degrees for 4 minutes. Serve in bed of salt to retain heat. Yield: 36 oysters.

Approx per oyster: Cal 25, Prot 2.1 gr,
T Fat 0.9 gr, Chol 11.3 mg, Carbo 2.4 gr,
Sod 82.7 mg, Potas 90.0 mg.

Ginny Farris
Lexington, KY

Scallops and Shells

2 c. sliced fresh mushrooms
½ c. chopped green onions
1 clove of garlic, minced
2 tbsp. butter

1 lb. fresh scallops
¼ c. dry white wine
¼ tsp. each thyme, salt
⅛ tsp. pepper
3 tbsp. melted butter
2 tbsp. flour
1 c. light cream
8 oz. large pasta shells, cooked
¼ c. chopped fresh parsley
2 tbsp. chopped pimento

Sauté mushrooms, green onions and garlic in 2 tablespoons butter in skillet. Add scallops, wine, thyme, salt and pepper. Simmer, covered, for 5 minutes. Blend 3 tablespoons butter and flour in saucepan over low heat. Cook for 1 minute, stirring constantly. Add cream gradually. Cook until thickened, stirring constantly. Drain scallop pan juices into cream sauce. Bring just to a boil. Add warm pasta; toss to coat well. Spoon onto serving platter. Mix parsley and pimento into scallop mixture. Spoon over pasta. Yield: 4 to 6 servings.

Photograph for this recipe on page 4.

Easy Shrimp and Rice

1 lb. peeled shrimp
1½ c. rice
1   10½-oz. can beef broth
1   4½-oz. can mushrooms, drained
1 onion, chopped
1 green pepper, chopped
¼ c. chopped green onion tops
¼ c. chopped parsley
1 stick margarine, melted
1 tsp. salt

Combine all ingredients in 8-cup electric rice cooker; mix well. Cook, covered, according to manufacturer's instructions. Let stand for 10 minutes before serving. Yield: 6 servings.

Approx per serving: Cal 424, Prot 24.5 gr,
T Fat 16.5 gr, Chol 122.8 mg, Carbo 42.9 gr,
Sod 979.8 mg, Potas 289.7 mg.

Dalta Gary
Acadiana H.S., Lafayette, LA

Shrimp Kabobs

2 lb. peeled shrimp
1 c. soy sauce
¼ c. lemon juice
1 med. onion, minced
8 cherry tomatoes
1 green pepper, cut into large pieces

Marinate shrimp in mixture of soy sauce, lemon juice and onion for 2 hours or longer. Drain. Thread shrimp, tomatoes and green pepper onto skewers. Broil for 10 minutes, turning frequently. Yield: 4 servings.

Approx per serving: Cal 285, Prot 56.0 gr, T Fat 2.7 gr, Chol 340.2 mg, Carbo 6.3 gr, Sod 324.2 mg, Potas 506.4 mg.

Ernestine Berry
Clearwater, FL

Shrimp Creole

1 c. sliced onions
½ c. chopped celery
1 clove of garlic, minced
3 tbsp. oil
1 tbsp. flour
1 tsp. salt
1 tsp. sugar
1 tbsp. chili powder
1 c. water
2 c. canned tomatoes
2 c. cooked peas
1 tbsp. vinegar
2 c. cooked shrimp
4 c. hot cooked rice

Sauté onions, celery and garlic in oil in skillet for 10 minutes. Add flour, salt, sugar and mixture of chili powder and ¼ cup water; mix well. Stir in ¾ cup water. Simmer for 15 minutes. Add tomatoes, peas, vinegar and shrimp. Heat to serving temperature. Pack rice into buttered 1-quart ring mold. Unmold on serving plate. Spoon shrimp mixture into center and around edge of rice ring. Yield: 6 servings.

Approx per serving: Cal 368, Prot 21.3 gr, T Fat 9.8 gr, Chol 76.8 mg, Carbo 48.0 gr, Sod 1183.2 mg, Potas 482.2 mg.

Marlene P. Thibault
Swanton, VT

Shrimp Divan

1 med. onion, finely chopped
½ c. chopped green pepper
1 apple, chopped
1 tbsp. curry powder
3 tbsp. margarine
3 tbsp. flour
2 c. low-fat milk
¼ c. lemon juice
1 can Cheddar cheese soup
3 c. cooked egg noodles
1 16-oz. package frozen broccoli, thawed
3 chicken breasts, cooked, sliced
1 lb. small shrimp, cooked

Sauté first 4 ingredients and salt to taste in margarine in saucepan. Stir in flour. Stir in mixture of milk, lemon juice and soup. Cook until heated through. Layer noodles, broccoli, chicken, shrimp and soup mixture in 9x13-inch baking dish. Bake at 350 degrees for 1 hour. Yield: 8 servings.

Approx per serving: Cal 306, Prot 32.6 gr, T Fat 7.6 gr, Chol 118.6 mg, Carbo 26.9 gr, Sod 201.0 mg, Potas 583.9 mg.

Kay Carthage
Raytown, MO

Shrimp in Wine Sauce

3 green onions, chopped
½ c. sliced fresh mushrooms
½ clove of garlic, chopped
2 tbsp. margarine
2 tbsp. flour
¾ c. low-fat milk
2 tbsp. white wine
¼ c. grated Swiss cheese
1½ c. cooked shrimp

Sauté vegetables in margarine in skillet; remove vegetables. Blend flour into pan juices. Stir in milk and wine gradually. Cook until thickened, stirring constantly. Stir in cheese until melted. Stir in sautéed vegetables and shrimp. Pour into buttered casserole. Bake at 350 degrees until bubbly. Yield: 4 servings.

Approx per serving: Cal 177, Prot 15.9 gr, T Fat 8.9 gr, Chol 80.9 mg, Carbo 7.4 gr, Sod 212.7 mg, Potas 210.0 mg.

Grant Hershey
North Platte, NM

Broccoli-Cheese Strata

20 oz. frozen chopped broccoli
6 slices whole wheat bread, trimmed
2 tbsp. margarine, softened
1½ c. shredded Cheddar cheese
5 eggs
2 tbsp. minced onion
¾ tsp. dry mustard
¾ tsp. Worcestershire sauce
¼ tsp. garlic powder
Pinch of cayenne pepper
1¾ c. milk
¾ c. shredded Cheddar cheese

Cook broccoli according to package directions; drain. Spread bread with margarine. Arrange in 7x11-inch baking dish. Layer 1½ cups cheese and broccoli over bread. Combine next 7 ingredients in bowl; mix well. Pour over layers. Chill, covered, for 8 hours to overnight. Bake, uncovered, for 30 minutes. Sprinkle with ¾ cup cheese. Bake for 10 minutes longer. Let stand for 5 minutes. Cut into squares. Yield: 6 servings.

Approx per serving: Cal 357, Prot 22.0 gr,
T Fat 25.2 gr, Chol 262.7 mg, Carbo 12.1 gr,
Sod 474.5 mg, Potas 442.3 mg.

Jean Turk
Wheaton North H.S., Wheaton, IL

Broccoli Puff with Sauce

10 oz. frozen chopped broccoli, cooked
6 eggs, separated
¼ tsp. each salt, pepper
½ tsp. cream of tartar
3 tbsp. flour
2 tbsp. margarine
2 c. low-fat milk
8 oz. American cheese, chopped
1 tbsp. Worcestershire sauce
1 tsp. dry mustard

Process broccoli, egg yolks, salt and pepper in blender container until smooth. Beat egg whites and cream of tartar until stiff. Fold in broccoli mixture. Spoon into ungreased 1-quart baking dish. Bake at 350 degrees for 20 minutes or until puffed and set. Blend flour and melted margarine in saucepan. Cook over medium heat until bubbly, stirring constantly. Stir in milk gradually. Cook over medium heat until thickened, stirring

constantly. Add remaining ingredients; mix well. Spoon hot broccoli puff onto serving plates. Top with cheese sauce. Yield: 6 servings.

Approx per serving: Cal 317, Prot 19.1 gr,
T Fat 19.8 gr, Chol 283.3 mg, Carbo 16.0 gr,
Sod 890.4 mg, Potas 429.7 mg.

Megan Herbert
Bowling Green, KY

Cheese Soufflé

⅓ c. flour
⅓ c. melted butter
1 tbsp. minced onion flakes
½ tsp. dry mustard
1½ c. milk
1 c. shredded Cheddar cheese
6 eggs, separated
¼ tsp. cream of tartar
2 tbsp. grated Parmesan cheese

Blend flour into ⅓ cup melted butter in saucepan. Add onion and mustard. Cook over medium heat until bubbly, stirring constantly. Stir in milk. Cook until thickened; remove from heat. Stir in Cheddar cheese until melted. Beat egg yolks in mixer bowl on high speed for 5 minutes. Stir a small amount of hot mixture into egg yolks. Stir egg yolks into hot mixture. Beat egg whites with cream of tartar in mixer bowl on high speed until stiff peaks form. Fold gently into egg yolks. Dust buttered soufflé dish with Parmesan cheese. Spoon in soufflé mixture. Bake at 350 degrees for 35 to 40 minutes or until puffy and lightly browned. Serve immediately. Yield: 6 servings.

Approx per serving: Cal 334, Prot 15.1 gr,
T Fat 26.2 gr, Chol 317.8 mg, Carbo 9.5 gr,
Sod 390.6 mg, Potas 192.8 mg.

Pearl Sorensen
Sonoma,CA

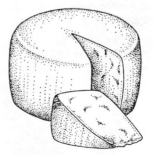

Swiss Custard

> 4 eggs, separated
> 1½ c. low-fat milk
> ⅛ tsp. nutmeg
> 1½ c. grated Swiss cheese
> Salt substitute to taste

Beat egg yolks with remaining ingredients in bowl. Fold in stiffly beaten egg whites gently. Pour into shallow 9-inch baking dish. Bake at 350 degrees for 30 minutes or until set. Serve immediately. Yield: 4 servings.

Approx per serving: Cal 278, Prot 21.2 gr,
T Fat 18.7 gr, Chol 299.0 mg, Carbo 5.6 gr,
Sod 408.9 mg, Potas 251.7 mg.

Ophelia Bjork
Raton, NM

Variety Custard

> 1 c. bread crumbs
> 1 c. sliced fresh mushrooms
> 1 c. sliced zucchini
> 1 c. fresh green beans
> 1 c. sliced carrots
> ½ c. chopped green pepper
> 1 c. chopped onion
> 2 tbsp. butter
> 4 eggs, well beaten
> 2 c. milk
> 1 tomato, sliced
> ¼ c. Parmesan cheese
> 2 tbsp. melted butter

Sprinkle bread crumbs in greased springform pan. Sauté next 6 vegetables in 2 tablespoons butter in skillet until tender-crisp. Spoon into prepared pan. Combine eggs and milk with salt, pepper and nutmeg to taste in bowl; mix well. Pour over vegetables. Top with sliced tomatoes, Parmesan cheese and 2 tablespoons melted butter. Bake at 375 degrees for 35 minutes or until set. Place on serving plate; remove side of pan. Cut into wedges. Yield: 8 servings.

Approx per serving: Cal 151, Prot 7.3 gr,
T Fat 8.3 gr, Chol 144.0 mg, Carbo 12.4 gr,
Sod 144.3 mg, Potas 331.1 mg.

Zelda Gorodetsky
Hoboken, NJ

Garden Lasagna

> 4 med. zucchini, coarsely chopped
> 1 lg. onion, chopped
> 1 med. green pepper, chopped
> 1 carrot, shredded
> ½ c. chopped celery
> 1 clove of garlic, minced
> 3 tbsp. oil
> 2 16-oz. cans stewed tomatoes
> 1 8-oz. can tomato sauce
> 1 6-oz. can tomato paste
> ¼ c. white wine
> 2 tbsp. chopped fresh parsley
> 2 tbsp. Italian seasoning
> 1 tsp. basil
> ½ tsp. seasoned salt
> ¼ tsp. freshly ground pepper
> 9 lasagna noodles
> 3 qt. water
> 1 16-oz. carton ricotta cheese
> 2 c. shredded mozzarella cheese
> 1 c. grated Parmesan cheese

Sauté first 6 ingredients in oil in skillet for 15 minutes. Add tomatoes, tomato sauce and paste, wine, parsley and seasonings; mix well. Simmer, covered, for 30 minutes. Simmer, uncovered, for 45 minutes longer or to desired consistency. Cook noodles in 3 quarts boiling water for 12 to 15 minutes or just until tender; drain. Layer ¼ of the sauce, 3 noodles, ⅓ of the ricotta cheese, ½ cup mozzarella cheese and ¼ cup Parmesan cheese in greased 9x13-inch baking dish. Repeat layers 2 times. Top with remaining sauce, mozzarella and Parmesan cheeses. Bake at 350 degrees for 35 to 40 minutes. Let stand for 5 minutes before serving. Yield: 8 servings.

Approx per serving: Cal 467, Prot 27.5 gr,
T Fat 20.3 gr, Chol 77.4 mg, Carbo 43.3 gr,
Sod 841.5 mg, Potas 1041.2 mg.

Traci Carnahan
Moore County, TX

Nutburgers

> 1 c. ground walnuts
> 1 c. shredded Cheddar cheese
> ½ c. bread crumbs
> ¼ c. wheat germ
> ½ c. chopped onion
> 2 tbsp. sesame seed

1 tbsp. chopped parsley
1 tsp. soy sauce
2 cloves of garlic, minced
1/4 tsp. thyme
2 tbsp. nonfat dry milk powder
4 eggs, lightly beaten
1/4 c. (about) oil

Combine first 11 ingredients in bowl; mix well. Add eggs; mix well. Shape into patties. Fry in oil in skillet over medium heat. Drain.
Yield: 5 servings.

Approx per serving: Cal 505, Prot 19.2 gr, T Fat 40.9 gr, Chol 225.8 mg, Carbo 18.6 gr, Sod 389.1 mg, Potas 359.1 mg.

Sandy Gill
Menominee, MI

Acapulco Omelet

1 sweet red pepper, chopped
1 green chili, finely chopped
1 onion, chopped
2 tbsp. margarine
12 eggs
2 tbsp. milk
1/2 tsp. salt
1/2 tsp. pepper
1/2 c. grated Cheddar cheese

Sauté red pepper, green chili and onion in margarine in skillet until tender. Remove with slotted spoon. Beat eggs with milk, salt and pepper in bowl. Pour into skillet; reduce heat. Cook over low heat until almost set, stirring occasionally. Top with sautéed vegetables and cheese. Fold omelet over; slide onto serving plate. Garnish with avocado slices and tomato wedges. Serve with hot corn tortillas.
Yield: 6 servings.

Approx per serving: Cal 241, Prot 16.1 gr, T Fat 18.0 gr, Chol 516.3 mg, Carbo 5.1 gr, Sod 106.2 mg, Potas 234.0 mg.

Susan Dillard
Florence, SC

Vegetable Omelets

1 small onion, chopped
1 zucchini, sliced
1 c. fresh sliced mushrooms

1/2 c. chopped green pepper
1 tomato, peeled, chopped
1/2 tsp. dried oregano leaves
1/2 tsp. salt
1/4 tsp. pepper
8 eggs
1/2 c. low-fat milk
1/2 c. shredded Swiss cheese

Sauté onion, zucchini, mushrooms and green pepper in nonstick skillet for 3 minutes or until tender. Add tomato, oregano and salt. Cook until heated through. Set aside. Beat eggs and milk in bowl until smooth. Heat nonstick skillet over medium heat until hot enough to sizzle a drop of water. Pour 1/2 cup egg mixture into skillet. Cook until set. Sprinkle 1/4 of the sautéed vegetables and 1/4 of the cheese over half the omelet; fold over to enclose filling. Place on serving plate. Repeat with remaining ingredients.
Yield: 4 servings.

Approx per serving: Cal 264, Prot 20.0 gr, T Fat 16.1 gr, Chol 521.0 mg, Carbo 9.9 gr, Sod 513.5 mg, Potas 543.8 mg.

Andy Estes
Sarasota, FL

Spinach Quiche

1   10-oz. package frozen chopped
    spinach, thawed
2 eggs, beaten
1/2 c. milk
1/2 c. grated Swiss cheese
2 tbsp. flour
1 sm. onion, chopped
1 c. sliced mushrooms
1 tsp. salt
Pepper to taste
1 unbaked 9-in. pie shell

Combine spinach, eggs, milk, cheese, flour, vegetables and seasonings in bowl; mix well. Spoon into pie shell. Bake at 350 degrees for 30 to 40 minutes or until set. Yield: 6 servings.

Approx per serving: Cal 261, Prot 9.6 gr, T Fat 15.9 gr, Chol 96.4 mg, Carbo 20.5 gr, Sod 667.4 mg, Potas 362.4 mg.

Mary Lu Buckman
Contra Costa, CA

Crustless Quiche Lorraine

2 tsp. butter, softened
2 tbsp. toasted wheat germ
4 slices lean bacon
3 tbsp. sliced green onions
5 eggs, beaten
1¼ c. milk
⅛ tsp. red pepper
1 c. shredded Swiss cheese

Spread butter in 9-inch quiche dish. Sprinkle with wheat germ. Fry bacon in skillet until crisp. Remove bacon and crumble. Drain skillet, reserving 1 tablespoon drippings. Sauté green onions in bacon drippings in skillet; drain. Beat eggs with milk and red pepper. Stir in crumbled bacon, green onions and cheese. Pour into prepared dish. Bake at 325 degrees for 30 minutes or until knife inserted in center comes out clean. Let stand for 5 to 10 minutes before serving. Yield: 6 servings.
Note: Nutritional information does not include bacon drippings.

Approx per serving: Cal 221, Prot 14.3 gr,
T Fat 16.0 gr, Chol 245.1 mg, Carbo 4.5 gr,
Sod 277.5 mg, Potas 184.8 mg.

Ruthie Andrews
Paul West Mid. Sch., East Point, GA

Mexican Quiche

¼ c. chopped jalapeño peppers
¾ c. grated Cheddar cheese
6 eggs
Lemon pepper to taste

Sprinkle jalapeño peppers in greased 8-inch pie plate. Top with Cheddar cheese. Combine eggs with salt and lemon pepper to taste in bowl. Beat with fork until smooth. Pour over cheese and peppers. Bake at 350 degrees for 25 minutes or until set. Cut into wedges. Yield: 6 servings.

Approx per serving: Cal 139, Prot 10.1 gr,
T Fat 10.3 gr, Chol 266.8 mg, Carbo 1.0 gr,
Sod 160.6 mg, Potas 84.7 mg.

Barbara Conmey
Farmington, ME

Confetti Spaghetti

8 oz. spaghetti
1 16-oz. package frozen mixed vegetables
1 tbsp. chopped onion
3 tbsp. margarine
1 tbsp. flour
⅛ tsp. nutmeg
¼ tsp. salt
1 c. low-fat milk
⅓ c. chopped red or green sweet pepper
1 2½-oz. jar sliced mushrooms, drained
¼ c. Parmesan cheese

Cook spaghetti and mixed vegetables according to package directions; drain. Sauté onion in margarine in saucepan. Stir in flour, nutmeg and salt. Add milk slowly. Bring to a boil. Cook for 1 minute or until thickened, stirring constantly; remove from heat. Stir in vegetables. Toss with spaghetti in serving dish. Sprinkle with cheese. Yield: 6 servings.

Approx per serving: Cal 275, Prot 10.2 gr,
T Fat 8.2 gr, Chol 6.3 mg, Carbo 40.5 gr,
Sod 250.5 mg, Potas 308.4 mg.

Cynthia Kolberg
Fairfield Jr.-Sr. H.S., Goshen, IN

Tomato-Egg Delight

1 10-oz. can tomato soup
4 eggs
¼ c. shredded American cheese
4 slices bread, toasted

Spoon soup into greased pie plate. Break eggs carefully into soup. Sprinkle with cheese. Bake at 400 degrees until eggs are set as desired. Place 1 piece toast on each serving plate. Lift 1 egg carefully onto each piece of toast. Spoon tomato sauce over top. Yield: 4 servings.

Approx per serving: Cal 239, Prot 11.8 gr,
T Fat 10.4 gr, Chol 260.0 mg, Carbo 24.6 gr,
Sod 889.3 mg, Potas 243.5 mg.

Hassie Hunter Green
Goshen H.S., Goshen, AL

Sauce for Fresh Asparagus

½ c. mayonnaise
½ c. cottage cheese
3 tbsp. chopped relish
2 tbsp. chopped green pepper
1 tsp. grated onion
1 tbsp. milk
1 tsp. vinegar

Combine all ingredients in bowl; mix well. Serve over fresh asparagus. Yield: 1 cup.

Photograph for this recipe on page 77.

Green Beans in Tomato Cups

6 med. tomatoes
1   9-oz. package frozen French-style
*   green beans, cooked*
1   4-oz. can sliced mushrooms, drained
6 tbsp. reduced-calorie Italian
*   salad dressing*
¼ c. sliced green onions
¼ tsp. salt
⅛ tsp. pepper

Cut thin slice from each tomato. Scoop out pulp, reserving shells. Sprinkle shells with salt and pepper to taste. Invert on paper towel-lined plate to drain. Chill in refrigerator. Combine remaining ingredients in bowl; mix well. Chill for 2 hours to overnight. Spoon into tomato cups. Serve on lettuce-lined plate. Yield: 6 servings.

Approx per serving: Cal 44, Prot 2.0 gr,
T Fat 1.0 gr, Chol 0.0 mg, Carbo 8.0 gr,
Sod 210.8 mg, Potas 338.5 mg.

Patricia Hancock
Kalona, IA

Tangy Green Beans

¾ lb. fresh green beans, cut crosswise
½ tsp. salt
1 med. onion, chopped
3 tbsp. butter
¼ tsp. each salt, dillweed
⅛ tsp. each pepper, savory
1 tbsp. wine vinegar
1 tbsp. butter

Cook beans with ½ teaspoon salt in a small amount of boiling water in saucepan until tender-crisp; drain. Sauté onion in 3 tablespoons butter in skillet for 3 to 5 minutes. Mix in beans. Cook until heated through, stirring occasionally. Add remaining ingredients. Toss lightly until butter is melted. Yield: 4 servings.

Approx per serving: Cal 142, Prot 2.2 gr,
T Fat 11.7 gr, Chol 35.5 mg, Carbo 9.1 gr,
Sod 549.0 mg, Potas 263.8 mg.

Sheryl Malone
Poway, CA

Aunt Lois's Microwave Hot Bean Salad

4 slices bacon
½ c. sugar
1 tbsp. cornstarch
⅔ c. vinegar
1 tsp. salt
¼ tsp. red pepper
1   16-oz. can cut green beans, drained
1   16-oz. can cut wax beans, drained
1   15-oz. can red kidney beans, drained
1 onion, thinly sliced

Fry bacon in skillet until crisp; crumble bacon and reserve. Combine bacon drippings, sugar and cornstarch in glass baking dish; mix well. Stir in vinegar, salt and red pepper. Microwave on High for 3 to 4 minutes or until thickened. Add beans and onion; mix well. Microwave for 6 minutes, stirring once. Let stand for 10 minutes. Sprinkle with bacon to serve. Yield: 15 servings.

Approx per serving: Cal 115, Prot 3.2 gr,
T Fat 4.4 gr, Chol 4.2 mg, Carbo 16.8 gr,
Sod 223.6 mg, Potas 181.9 mg.

Dalta Gary
Acadiana H.S., Lafayette, LA

Beets in Orange Sauce

1   16-oz. can sliced beets
Ginger and cinnamon to taste
1 tbsp. frozen orange juice concentrate

Combine undrained beets with seasonings and orange juice concentrate in saucepan. Cook until heated through. Yield: 4 servings.

Approx per serving: Cal 39, Prot 1.0 gr,
T Fat 0.1 gr, Chol 0.0 mg, Carbo 9.2 gr,
Sod 200.7 mg, Potas 171.9 mg.

Mary Ann Madar
Elizabeth Township, PA

Broccoli Casserole

1 lg. bunch broccoli, trimmed
1 can mushroom soup
2 eggs
1 tbsp. onion flakes
½ c. salad dressing
1 c. grated Cheddar cheese
1 stick margarine, melted
½ stack crackers, crushed

Cook broccoli in a small amount of water in saucepan until tender-crisp; drain. Combine soup, eggs, onion flakes, salad dressing, cheese and salt and pepper to taste in bowl; mix well. Stir in broccoli. Spoon into 3-quart casserole. Mix margarine and cracker crumbs. Sprinkle over broccoli mixture. Bake at 350 degrees for 30 minutes. Yield: 6 servings.

Approx per serving: Cal 445, Prot 11.5 gr,
T Fat 37.6 gr, Chol 116.9 mg, Carbo 18.1 gr,
Sod 959.0 mg, Potas 382.7 mg.

Marilyn Mancewicz
Ottawa Hills H.S., Grand Rapids, MI

Linda's Broccoli Casserole

2   10-oz. packages frozen broccoli,
* thawed*
1 sm. onion, chopped
2 eggs
½ c. milk
½ c. shredded mozzarella cheese
1 c. bread crumbs
½ c. melted butter

Arrange broccoli in 10x10-inch baking dish. Combine onion, eggs, milk and cheese in bowl; mix well. Pour over broccoli. Sprinkle bread crumbs over top. Drizzle with butter. Bake at 350 degrees for 40 to 50 minutes or until broccoli is tender. Yield: 8 servings.

Approx per serving: Cal 198, Prot 6.8 gr,
T Fat 15.7 gr, Chol 107.3 mg, Carbo 9.3 gr,
Sod 241.2 mg, Potas 241.9 mg.

Linda K. Turner
Northeastern H.S., Richmond, IN

Broccoli-Rice Casserole

1 tbsp. chopped onion
¼ c. butter
1   10-oz. package frozen chopped
* broccoli*
1 c. minute rice
1 can cream of celery soup
1½ c. shredded Cheddar cheese

Sauté onion in butter in skillet. Add broccoli. Heat until broccoli is thawed, stirring frequently. Add remaining ingredients; mix well. Spoon into 1½-quart casserole. Bake at 350 degrees for 30 minutes. Yield: 4 servings.

Approx per serving: Cal 434, Prot 15.8 gr,
T Fat 28.5 gr, Chol 81.9 mg, Carbo 30.0 gr,
Sod 1042.4 mg, Potas 280.2 mg.

Fran Heckman
Waupaca H.S., Waupaca, WI

Confetti Cabbage

1 med. head red cabbage, sliced
2 tbsp. red wine vinegar
1 tsp. sugar
½ tsp. salt
3 tbsp. oil
10 oz. frozen peas
1 med. head green cabbage, sliced
1 med. onion, chopped
½ tsp. each caraway seed, salt
3 tbsp. oil

Combine red cabbage, vinegar, sugar, ½ teaspoon salt and 3 tablespoons oil in saucepan. Cook for 25 minutes or until cabbage is tender, stirring occasionally. Stir in peas. Cook until heated through. Combine green cabbage with remaining ingredients in saucepan. Cook for 20 minutes or until cabbage and onion are tender, stirring occasionally. Arrange on serving platter. Yield: 10 servings.

Approx per serving: Cal 262, Prot 12.0 gr,
T Fat 9.3 gr, Chol 0.0 mg, Carbo 41.5 gr,
Sod 387.2 mg, Potas 1492.7 mg.

Sandra McDow
Fresno, CA

Microwave Red Cabbage

5 c. shredded red cabbage
1 c. chopped peeled apple
½ c. raisins
½ c. chopped onion
1 tbsp. sugar
2 tbsp. butter
2 tbsp. cider vinegar
¼ tsp. cinnamon
1 tsp. salt
¼ tsp. pepper
¼ c. water

Combine all ingredients in glass baking dish. Microwave, covered, on High for 12 minutes or until cabbage is tender-crisp, stirring once. Let stand for several minutes. Yield: 6 servings.

Approx per serving: Cal 113, Prot 1.8 gr,
T Fat 4.1 gr, Chol 11.8 mg, Carbo 19.9 gr,
Sod 422.1 mg, Potas 299.9 mg.

Nola Parker
Memphis, TN

Stir-Fry Cabbage

1 onion, sliced
1 c. sliced celery
4 c. shredded cabbage
1 tbsp. oil
½ c. chicken broth
2 tbsp. soy sauce

Stir-fry onion, celery and cabbage in oil in skillet for 1 minute. Add broth and soy sauce. Cook over high heat until cabbage is tender-crisp, stirring constantly. Yield: 8 servings.

Approx per serving: Cal 40, Prot 1.1 gr,
T Fat 2.1 gr, Chol 0.0 mg, Carbo 5.1 gr,
Sod 480.3 mg, Potas 206.2 mg.

Nelda Loper
Shawnee, OK

Glazed Carrots

2 lg. carrots, cut into julienne strips
1 navel orange, sectioned
Pinch of cinnamon
1 tsp. butter flavoring
Sugar to taste

Combine carrots with a small amount of water in saucepan. Cook for 10 minutes or until tender; drain. Add remaining ingredients; mix gently. Cook until heated through. Yield: 2 servings.
Note: Nutritional information does not include sugar to taste.

Approx per serving: Cal 66, Prot 1.7 gr,
T Fat 0.2 gr, Chol 0.0 mg, Carbo 15.9 gr,
Sod 34.6 mg, Potas 381.7 mg.

Martha Gordon
Newport News, VA

Carrot and Apple Casserole

6 tbsp. sugar
2 tbsp. flour
¼ tsp. salt
5 apples, thinly sliced
2 c. sliced peeled carrots
¾ c. orange juice

Combine first 3 ingredients in bowl. Alternate layers of apples and carrots in baking dish until all ingredients are used, sprinkling each layer with sugar mixture. Pour orange juice over layers. Bake at 350 degrees for 20 to 30 minutes or just until carrots are tender. Yield: 4 servings.

Approx per serving: Cal 281, Prot 1.8 gr,
T Fat 1.8 gr, Chol 0.0 mg, Carbo 69.4 gr,
Sod 163.5 mg, Potas 579.2 mg.

Candy Overton
Big Spring, TX

Cauliflower-Broccoli Medley

Flowerets of 1 head cauliflower
Flowerets of 1 bunch broccoli
2 sm. onions, sliced into rings
½ c. mayonnaise
⅓ c. oil
⅓ c. vinegar
¼ c. sugar
½ tsp. salt
¼ tsp. pepper
2 slices crisp-fried bacon, crumbled

Mix vegetables in bowl. Combine mayonnaise, oil, vinegar, sugar and seasonings in small bowl. Pour over vegetables; mix well. Chill for several hours to overnight. Sprinkle with bacon at serving time. Yield: 6 servings.

Approx per serving: Cal 363, Prot 7.9 gr,
T Fat 28.8 gr, Chol 15.3 mg, Carbo 23.9 gr,
Sod 347.6 mg, Potas 781.9 mg.

Nadine H. Shipp
Fyffe Sch., Rainsville, AL

Creamy Corn Italiano

½ c. diagonally sliced green onions
2 med. cloves of garlic, minced
½ tsp. oregano
¼ tsp. basil
2 tbsp. butter
2 16-oz. cans whole kernel corn, drained
½ c. chopped drained tomatoes
1 can cream of celery soup
¼ c. milk

Sauté first 4 ingredients in butter in 2-quart saucepan. Add remaining ingredients; mix well. Heat to serving temperature, stirring occasionally. Yield: 6 servings.

Approx per serving: Cal 175, Prot 4.2 gr,
T Fat 7.2 gr, Chol 16.2 mg, Carbo 27.4 gr,
Sod 707.6 mg, Potas 215.9 mg.

Carolyn Thompson
Redwater H.S., Texarkana, TX

Special Baked Corn

1 16-oz. can whole kernel corn, drained
1 16-oz. can cream-style corn
1 med. green pepper, chopped
1 sm. onion, chopped
1 2-oz. jar chopped pimento
1 egg, beaten
1 c. grated Cheddar cheese
1 c. cracker crumbs
¼ c. melted margarine

Combine all ingredients with salt and pepper to taste in bowl; mix well. Spoon into greased 8x10-inch baking dish. Bake at 350 degrees for 45 minutes or until golden. Yield: 8 servings.

Approx per serving: Cal 253, Prot 8.0 gr,
T Fat 12.9 gr, Chol 45.6 mg, Carbo 29.8 gr,
Sod 526.5 mg, Potas 206.5 mg.

Patricia B. Foster
Scott Mid. Sch., Scott, LA

Corn and Broccoli Dish

½ c. butter
1 12-oz. can whole kernel corn
with peppers, drained
1 10-oz. package frozen broccoli,
thawed, drained
¼ tsp. minced garlic
½ tsp. salt
½ tsp. pepper

Melt butter in 2-quart saucepan. Add remaining ingredients; mix well. Simmer for 8 to 10 minutes, stirring occasionally. Yield: 4 servings.

Approx per serving: Cal 276, Prot 4.1 gr,
T Fat 23.7 gr, Chol 71.0 mg, Carbo 16.2 gr,
Sod 704.9 mg, Potas 239.7 mg.

Gail Dixon
Lyons Jr. H.S., Vidalia, GA

Eggplant Casserole

2 med. eggplant, peeled, chopped
4 tomatoes, chopped
2 green peppers, thinly sliced
2 onions, sliced into rings
½ tsp. salt
½ tsp. pepper
2 tbsp. oil
1 c. bread crumbs
2 tbsp. tapioca
8 oz. Cheddar cheese, grated

Layer half the eggplant, tomatoes, green peppers, onions, salt, pepper, oil and all the bread crumbs in greased shallow 2-quart baking dish. Repeat layers. Sprinkle tapioca over layers. Bake, covered with foil, for 1 hour. Top with cheese. Bake, uncovered, until cheese melts. Yield: 8 servings.

Approx per serving: Cal 232, Prot 11.0 gr,
T Fat 13.1 gr, Chol 28.3 mg, Carbo 20.1 gr,
Sod 378.2 mg, Potas 512.1 mg.

Myrle L. Swicegood
Oconee, SC

Leek Pie

6 potatoes, peeled, sliced
4 leeks, sliced ½ in. thick
1 tbsp. cornstarch
1½ c. milk
¼ c. melted butter
¼ tsp. salt
⅛ tsp. nutmeg
¼ tsp. pepper
¼ c. shredded Cheddar cheese

Layer half the potatoes, all the leeks and remaining potatoes in greased 2½-quart casserole. Blend cornstarch with 2 tablespoons milk in saucepan. Add remaining milk and butter. Bring to a boil, stirring constantly; remove from heat. Stir in seasonings and cheese. Pour over vegetables. Bake, covered, at 350 degrees for 45 minutes. Garnish with additional cheese. Broil until browned if desired. Yield: 6 servings.

Approx per serving: Cal 283, Prot 7.5 gr,
T Fat 11.5 gr, Chol 36.9 mg, Carbo 38.5 gr,
Sod 252.3 mg, Potas 903.2 mg.

Maureen Munro Dillard
Los Angeles, CA

Baked Onion Rings

2 egg whites
½ tsp. salt
⅛ tsp. pepper
½ lg. sweet yellow onion, cut into rings
⅓ c. dry bread crumbs

Combine egg whites, salt and pepper in bowl; mix well. Dip onion rings into egg mixture. Coat with bread crumbs. Place in single layer on baking sheet. Bake at 450 degrees for 10 minutes or until golden. Yield: 4 servings.

Approx per serving: Cal 36, Prot 2.7 gr,
T Fat 0.2 gr, Chol 0.2 mg, Carbo 5.6 gr,
Sod 328.6 mg, Potas 63.8 mg.

Tina Rollins
Boone, NC

Stuffed Onions

8 lg. onions, peeled
¾ c. Brazil nuts, finely chopped
1½ c. cooked wild rice

2 tbsp. lemon juice
1 egg, slightly beaten
1 tsp. salt
⅛ tsp. pepper
¼ c. bread crumbs
1 tbsp. butter
Paprika

Remove slice from tops of onions. Cook onions in boiling water in saucepan for 30 minutes or until almost tender. Scoop out centers leaving ¼ to ½-inch thick shells. Place in shallow baking dish. Combine nuts, wild rice, lemon juice, egg, salt and pepper in bowl; mix well. Spoon into onion shells. Mix crumbs with butter. Sprinkle crumbs and paprika over onions. Add a small amount of water to baking dish. Bake at 375 degrees for 30 minutes or until browned.
Yield: 8 servings.

Approx per serving: Cal 264, Prot 8.3 gr,
T Fat 14.3 gr, Chol 36.1 mg, Carbo 28.5 gr,
Sod 386.5 mg, Potas 473.8 mg.

Linda Rice
Berkeley, CA

Dixie Peas

⅓ c. chopped onion
1½ c. thinly sliced celery
3 tbsp. margarine
1   10-oz. package frozen peas
¾ tsp. salt
Dash of pepper
Dash of thyme
2 tbsp. water
½ tsp. Worcestershire sauce
2 tbsp. chopped pimento
1 tsp. chopped parsley

Sauté onion and celery in margarine in saucepan for 5 minutes. Add peas, salt, pepper, thyme and water. Simmer, covered, for 6 to 8 minutes or just until peas are tender. Stir in Worcestershire sauce, pimento and parsley. Yield: 6 servings.

Approx per serving: Cal 95, Prot 3.1 gr,
T Fat 6.1 gr, Chol 0.0 mg, Carbo 8.2 gr,
Sod 439.3 mg, Potas 200.5 mg.

Sandra Isaac
Montgomery, AL

Hopping John Peas

4 slices bacon, chopped
¼ c. chopped onion
¼ c. rice
1   20-oz. can black-eyed peas
1 c. water

Cook bacon in skillet until crisp. Add onion, rice, peas and water. Simmer, covered, until rice is tender. Yield: 8 servings.

Approx per serving: Cal 148, Prot 5.0 gr,
T Fat 8.1 gr, Chol 7.9 mg, Carbo 14.0 gr,
Sod 245.2 mg, Potas 277.9 mg.

Natalie McBee
Cleveland, OH

Savory Baked Potatoes

6 baking potatoes
6   ¼-in. thick onion slices
6 slices green pepper
1 tsp. garlic salt
Pepper to taste
6 tbsp. chicken broth

Cut each potato in half lengthwise. Place onion and green pepper slice between halves. Sprinkle with garlic salt and pepper to taste. Place on foil. Drizzle 1 tablespoon broth over each. Seal foil tightly. Bake at 425 degrees for 1 hour or until tender. Yield: 6 servings.

Approx per serving: Cal 166, Prot 5.9 gr,
T Fat 0.4 gr, Chol 4.6 mg, Carbo 36.0 gr,
Sod 521.5 mg, Potas 866.9 mg.

Margie Hendricks
Little Rock, AR

Creamy Scalloped Potatoes

6 med. potatoes, sliced
3 tbsp. margarine
1 sm. onion, finely chopped
1 tsp. salt
¼ tsp. pepper
3 tbsp. flour
2½ c. milk

Layer potatoes, margarine, onion, seasonings and flour alternately in baking dish until all ingredients are used. Pour milk over layers.

Bake at 425 degrees for 1 hour or until potatoes are tender. Yield: 12 servings.

Approx per serving: Cal 140, Prot 4.1 gr,
T Fat 4.8 gr, Chol 7.1 mg, Carbo 20.6 gr,
Sod 241.7 mg, Potas 468.6 mg.

Ruth Landry
Idabel, OK

Foiled Potatoes

4 lg. baking potatoes
1 tbsp. reduced-calorie margarine
1 tbsp. Parmesan cheese
1 tsp. chopped chives
½ tsp. pepper
⅛ tsp. garlic powder
1 med. onion, thinly sliced

Cut potatoes into halves crosswise. Cut ¾-inch slices to but not through bottom. Mix next 5 ingredients in bowl. Place a small amount of mixture and 1 onion slice in each slit. Wrap in foil. Bake at 400 degrees for 45 minutes or until tender. Yield: 4 servings.

Approx per serving: Cal 162, Prot 5.1 gr,
T Fat 0.7 gr, Chol 1.7 mg, Carbo 34.9 gr,
Sod 21.9 mg, Potas 815.8 mg.

Macie Phillips
Andover, MA

Swiss-Potato Bake

½ c. low-fat milk
2 tbsp. chopped pimento
2 tbsp. chopped parsley
2 med. potatoes, peeled, thinly sliced
3 med. onions, thinly sliced
½ c. shredded Swiss cheese

Combine milk, pimento and parsley in bowl. Layer potatoes, onions, salt and pepper to taste and milk mixture ½ at a time in greased 6x10-inch baking dish. Bake, covered, at 350 degrees for 1 hour or until potatoes are tender. Top with cheese. Bake until cheese melts.
Yield: 6 servings.

Approx per serving: Cal 120, Prot 5.7 gr,
T Fat 3.0 gr, Chol 10.3 mg, Carbo 18.4 gr,
Sod 87.9 mg, Potas 426.3 mg.

Marietta O'Roark
Princeton, NJ

Mashed Rutabagas and Potatoes

1 tbsp. sugar
4 c. chopped peeled rutabagas
3 c. chopped peeled potatoes
1 chicken bouillon cube
2 c. boiling water
1/4 tsp. pepper
1 c. grated Cheddar cheese
2 tbsp. finely chopped onion

Sprinkle sugar over rutabagas and potatoes in saucepan. Dissolve bouillon cube in boiling water. Pour over vegetables. Bring to a boil; reduce heat. Simmer until vegetables are tender. Drain and mash vegetables. Add pepper, cheese and onion. Beat until fluffy. Spoon into serving dish. Yield: 6 servings.

*Approx per serving: Cal 185, Prot 7.5 gr,
T Fat 6.3 gr, Chol 19.2 mg, Carbo 25.8 gr,
Sod 288.7 mg, Potas 550.3 mg.*

Pearl McCauley
Delta, MI

Spinach Casserole

1   16-oz. carton cottage cheese
8 oz. sharp Cheddar cheese, shredded
1   10-oz. package frozen chopped
    spinach, thawed, drained
3 eggs, beaten

Combine all ingredients in bowl; mix well. Spoon into buttered casserole. Bake at 350 degrees for 45 minutes. Yield: 8 servings.

*Approx per serving: Cal 217, Prot 19.2 gr,
T Fat 14.1 gr, Chol 135.3 mg, Carbo 4.0 gr,
Sod 382.2 mg, Potas 225.3 mg.*

Talley West
Pickens, SC

Microwave Spinach Casserole

3 eggs
1 c. stuffing mix
1 tbsp. stuffing mix seasoning
1   10-oz. package frozen chopped
    spinach, thawed
1 lg. onion, chopped
1/4 c. margarine, softened
1/4 c. Parmesan cheese

1/4 tsp. garlic powder
1/4 tsp. thyme
1/2 tsp. pepper

Beat eggs in mixer bowl. Add stuffing mix, stuffing mix seasoning and remaining ingredients; mix well. Spoon into 1-quart glass baking dish sprayed with nonstick cooking spray. Microwave on Medium-High for 10 minutes. Let stand for 5 minutes before serving.
Yield: 6 servings.
Note: Nutritional information does not include stuffing mix or stuffing mix seasoning.

*Approx per serving: Cal 151, Prot 7.0 gr,
T Fat 12.0 gr, Chol 131.1 mg, Carbo 4.7 gr,
Sod 190.1 mg, Potas 253.9 mg.*

Sue Lawson
Haworth Sch., Idabel, OK

Spinach Roll

2   10-oz. packages chopped spinach,
    cooked, drained
6 tbsp. melted butter
Dash of nutmeg
3 eggs, separated
1 c. chopped onion
15 fresh mushrooms, sliced
3 tbsp. butter
1/2 tsp. salt
Basil, parsley and pepper to taste
1 c. grated Cheddar cheese

Combine spinach, melted butter, nutmeg and beaten egg yolks in bowl; mix well. Fold in stiffly beaten egg whites. Spread over greased baking sheet. Bake at 350 degrees for 20 minutes. Loosen with spatula; invert onto foil. Sauté onion and mushrooms in 3 tablespoons butter in saucepan. Stir in salt, basil, parsley and pepper. Spread over spinach. Sprinkle with 3/4 cup cheese. Roll as for jelly roll. Place on baking sheet. Top with remaining cheese. Bake for 10 to 20 minutes longer or until cheese melts.
Yield: 12 servings.

*Approx per serving: Cal 154, Prot 6.0 gr,
T Fat 13.3 gr, Chol 99.1 mg, Carbo 3.9 gr,
Sod 305.3 mg, Potas 264.5 mg.*

Kim Terry
Amarillo, TX

Spinach-Mushroom Casserole

1   8-oz. can mushrooms
2 tbsp. butter
2   10-oz. packages frozen spinach,
    thawed, drained
2 eggs
1   8-oz. carton sour cream
¼ c. Parmesan cheese
⅛ tsp. nutmeg
½ tsp. salt
½ tsp. pepper
¼ c. Parmesan cheese

Sauté mushrooms in butter in skillet for 10 minutes. Add spinach, eggs, sour cream, ¼ cup Parmesan cheese and seasonings; mix well. Spoon into greased 2-quart baking dish. Sprinkle with remaining Parmesan cheese. Bake at 350 degrees for 45 minutes. Yield: 6 servings.

*Approx per serving: Cal 211, Prot 10.3 gr,
T Fat 16.7 gr, Chol 122.4 mg, Carbo 6.5 gr,
Sod 391.5 mg, Potas 427.7 mg.*

*Barbara E. Hancock
Highland Springs H.S., Richmond, VA*

Acorn Squash

3 med. acorn squash
1   15-oz. can pineapple chunks, drained
¼ c. butter, melted
⅓ c. packed brown sugar
½ c. chopped walnuts

Cut squash into halves; remove seed. Place cut side down in large baking dish. Bake at 400 degrees for 35 minutes; turn squash. Mix remaining ingredients in bowl. Spoon into centers of squash. Bake for 10 minutes longer. Yield: 6 servings.

*Approx per serving: Cal 338, Prot 5.5 gr,
T Fat 13.9 gr, Chol 23.6 mg, Carbo 54.3 gr,
Sod 100.1 mg, Potas 1016.3 mg.*

*Sharon Lynne Willis
Sequatchie, TN*

Squash Cubes in Sour Cream

3 tbsp. butter
1 med. onion, sliced
4 c. chopped Hubbard squash
½ tsp. salt
¼ tsp. pepper
1 tbsp. sugar
¼ c. water
1 c. sour cream
½ tsp. dillseed

Melt butter in skillet. Layer onion and squash evenly in skillet. Sprinkle with salt, pepper and sugar. Add water. Cook, tightly covered, over low heat for 10 minutes. Turn squash carefully. Cook for 10 minutes longer. Heat sour cream in double boiler, stirring constantly. Invert squash onto warm platter. Spoon sour cream over top. Sprinkle with dillseed. Yield: 6 servings.

*Approx per serving: Cal 196, Prot 3.3 gr,
T Fat 14.3 gr, Chol 34.6 mg, Carbo 16.3 gr,
Sod 449.5 mg, Potas 328.4 mg.*

*Verda Griner
Los Angeles, CA*

Summer Squash Casserole

2 tbsp. rice
1 lb. squash, cut into 1-in. pieces
1 lg. onion, chopped
2 lg. tomatoes, chopped
1 lg. green pepper, chopped
1 stalk celery, finely chopped
½ tsp. salt
Pepper to taste
1 tbsp. brown sugar
1 tbsp. butter

Sprinkle rice into large greased casserole. Layer vegetables over rice. Sprinkle with seasonings and brown sugar. Dot with butter. Bake, covered, at 350 degrees for 1½ hours, stirring after 1 hour. Yield: 4 servings.

*Approx per serving: Cal 116, Prot 3.3 gr,
T Fat 3.2 gr, Chol 8.9 mg, Carbo 20.5 gr,
Sod 325.0 mg, Potas 536.6 mg.*

*Gail Bierstedt
Seguin, WA*

Baked Tomato Halves

6 med. tomatoes
2 tbsp. prepared brown mustard
3 c. fresh bread cubes
¼ c. melted butter
¼ tsp. salt
⅛ tsp. pepper
1½ tsp. Worcestershire sauce
⅛ tsp. Tabasco sauce

Cut tomatoes into halves crosswise. Arrange in shallow baking dish. Spread each half with ½ teaspoon mustard. Toss bread cubes with butter and seasonings in bowl. Spoon 2 tablespoons bread mixture onto each tomato half. Bake at 375 degrees for 20 to 25 minutes or until tomatoes are tender and topping is golden. Yield: 6 servings.

Approx per serving: Cal 176, Prot 4.1 gr,
T Fat 9.0 gr, Chol 24.5 mg, Carbo 20.7 gr,
Sod 405.4 mg, Potas 348.3 mg.

> Pat Stanley
> Carmichael, CA

Parmesan Zucchini

3 c. sliced zucchini
4 c. canned tomatoes
1 med. onion, sliced
½ tsp. each Italian seasoning, salt
Pepper to taste
¼ c. Parmesan cheese

Layer zucchini, tomatoes and onion alternately in greased 2-quart casserole until all ingredients are used. Sprinkle with seasonings. Top with Parmesan cheese. Bake at 350 degrees for 1 hour. Yield: 6 servings.

Approx per serving: Cal 72, Prot 4.5 gr,
T Fat 1.5 gr, Chol 4.7 mg, Carbo 11.2 gr,
Sod 425.5 mg, Potas 520.7 mg.

> Jill Prichard
> Erie, PA

Zucchini Casserole

8 zucchini, sliced
1 stick margarine
5 yellow onions, sliced
6 tomatoes, sliced

1 c. grated sharp cheese
½ c. bread crumbs
2 tbsp. butter

Sauté a small amount of zucchini at a time in margarine in skillet. Sauté onions in skillet. Layer zucchini, onions, tomatoes and cheese in 9x13-inch baking dish until all ingredients are used. Sprinkle with bread crumbs. Dot with butter. Bake at 350 degrees for 25 to 30 minutes or until bubbly. Yield: 8 servings.

Approx per serving: Cal 287, Prot 8.9 gr,
T Fat 19.5 gr, Chol 23.0 mg, Carbo 22.8 gr,
Sod 307.4 mg, Potas 804.9 mg.

> Ella Jo Adams
> Allen H.S., Allen, TX

Vegetable Variety Casserole

1 16-oz. can cut green beans, drained
1 16-oz. can mixed vegetables, drained
1 can cream of chicken soup
½ c. crushed potato chips
2 tbsp. butter

Combine green beans, mixed vegetables and soup in bowl; mix well. Spoon into greased 8x8-inch baking dish. Sprinkle with potato chips. Dot with butter. Bake at 350 degrees for 45 minutes. Yield: 4 servings.

Approx per serving: Cal 271, Prot 6.6 gr,
T Fat 15.6 gr, Chol 23.7 mg, Carbo 29.2 gr,
Sod 1244.9 mg, Potas 495.5 mg.

> Emily Lewis
> Capitol Hill H.S., Oklahoma City, OK

Microwave Herbed Vegetable Medley

1 1 to 1½-lb. cauliflower, trimmed
1 1-lb. bunch broccoli, cut into
*   short spears*
4 carrots, cut diagonally into ¼-in. slices
½ c. butter
¼ tsp. freshly ground pepper
1½ tsp. chopped parsley
1½ tsp. chopped fresh chives

Place cauliflower in center or 15-inch plate. Arrange broccoli and carrots around cauliflower. Cover with vented plastic wrap. Microwave on High for 15 to 18 minutes or until tender-crisp.

Let stand, covered, for several minutes; drain. Combine remaining ingredients in small glass bowl. Microwave on High for 1 to 1½ minutes or until melted; mix well. Drizzle over vegetables. Yield: 6 servings.

Approx per serving: Cal 211, Prot 6.4 gr, T Fat 15.9 gr, Chol 47.3 mg, Carbo 15.1 gr, Sod 235.5 mg, Potas 791.4 mg.

Sue P. Culpepper
Gulfport H.S., Gulfport, MS

Vegetable Pie with Potato Crust

3 med. potatoes, peeled, cooked
¼ c. butter
½ c. each chopped onion, green pepper
2 c. frozen mixed vegetables
1 c. shredded Cheddar cheese
1 egg, beaten
1 5⅓-oz. can evaporated milk
1 tsp. salt
½ tsp. pepper
2 tbsp. toasted wheat germ

Mash enough potatoes to measure 1⅓ cups. Mix with butter. Spread over bottom and side of greased 9-inch pie plate to form crust. Combine vegetables in prepared pie plate. Sprinkle with cheese. Beat egg with evaporated milk, salt and pepper in bowl. Pour over vegetables and cheese. Sprinkle wheat germ over top. Bake at 375 degrees for 40 to 50 minutes or until set. Yield: 8 servings.

Approx per serving: Cal 258, Prot 10.4 gr, T Fat 13.2 gr, Chol 69.9 mg, Carbo 26.3 gr, Sod 514.3 mg, Potas 570.0 mg.

Mildred Oglesby
Dalton, GA

Vegetable Medley

2 lb. zucchini, sliced ¼ in. thick
½ c. each chopped green pepper, onion
1 clove of garlic, minced
1 tsp. salt
¼ tsp. pepper
4 med. tomatoes, peeled, chopped
2 tbsp. butter
1 tsp. parsley flakes

Mix first 6 ingredients in Crock•Pot. Top with tomatoes, butter and parsley. Cook on Low for 3 to 4 hours or until zucchini is tender. Yield: 6 servings.

Approx per serving: Cal 83, Prot 2.9 gr, T Fat 4.2 gr, Chol 11.0 mg, Carbo 10.5 gr, Sod 408.0 mg, Potas 515.0 mg.

Rose H. Alexander
Kanawha, WV

Green and Gold Casserole

1 lb. fresh zucchini, sliced
1½ c. small-curd cottage cheese
2 tbsp. sour cream
2 tbsp. flour
1 tbsp. lemon juice
¼ tsp. pepper
¼ tsp. Tabasco sauce
2 eggs
1 16-oz. can corn, drained
1 to 2 tbsp. chopped green chilies
½ c. grated Cheddar cheese
½ c. seasoned bread crumbs

Cook zucchini in a small amount of boiling salted water in saucepan until tender-crisp; drain. Combine cottage cheese, sour cream, flour, lemon juice, seasonings and eggs in blender container. Process until smooth. Fold zucchini, corn, chilies and cottage cheese mixture together in bowl. Spoon into 1½-quart casserole. Top with Cheddar cheese and crumbs. Bake at 350 degrees for 45 minutes. Yield: 6 servings.

Approx per serving: Cal 224, Prot 16.0 gr, T Fat 9.0 gr, Chol 134.7 mg, Carbo 7.1 gr, Sod 388.4 mg, Potas 277.3 mg.

Vivienne Webber
Van Nuys, CA

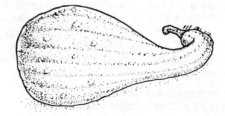

Ratatouille

6 c. cubed eggplant
3 c. sliced zucchini
¾ c. chopped green pepper
1 c. chopped celery
¾ c. chopped onion
1 clove of garlic, chopped
⅓ c. oil
3 tomatoes, chopped
1 4-oz. can mushrooms, drained
½ c. sliced green olives
1½ tsp. salt
1 tsp. oregano
½ tsp. thyme
¼ tsp. pepper

Sauté eggplant, zucchini, green pepper, celery, onion and garlic in oil in 4-quart saucepan for 7 to 10 minutes or until vegetables are tender-crisp, stirring occasionally. Add tomatoes, mushrooms, olives and seasonings. Simmer, covered, until heated through. Yield: 8 servings.

Approx per serving: Cal 134, Prot 2.7 gr,
T Fat 9.9 gr, Chol 0.0 mg, Carbo 10.9 gr,
Sod 703.9 mg, Potas 473.0 mg.

Carla Schroer
Sarasota, FL

Stir-Fry Ratatouille

1 clove of garlic, minced
1 tbsp. oil
2½ c. thinly sliced zucchini
1 sm. onion, thinly sliced
1 c. green pepper strips
1 c. chopped tomato
1 tbsp. chopped parsley
1 tbsp. vinegar
¼ tsp. crushed dried basil
¼ tsp. crushed dried oregano

Stir-fry garlic in oil in skillet for 15 seconds. Add garlic. Stir-fry for 15 seconds. Add zucchini and onion. Stir-fry for 2 minutes. Add green pepper. Stir-fry for 2 minutes. Add tomato and remaining ingredients. Simmer, covered, for 1 minute. Spoon into serving dish. Yield: 8 servings.

Approx per serving: Cal 41, Prot 1.6 gr,
T Fat 1.8 gr, Chol 0.0 mg, Carbo 5.8 gr,
Sod 5.0 mg, Potas 265.4 mg.

Christie Campbell
Ft. Lauderdale, FL

Three-Vegetable Casserole

1 10-oz. package frozen chopped
    broccoli
1 10-oz. package frozen French-style
    green beans
1 10-oz. package frozen cauliflower
2 cans cream of mushroom soup
1 c. shredded Cheddar cheese
½ c. croutons

Cook vegetables in a small amount of water in saucepan until tender-crisp. Drain well. Add soup and ¾ cup cheese; mix well. Spoon into greased 9x13-inch baking dish. Bake at 350 degrees for 30 minutes. Sprinkle with remaining ¼ cup cheese and croutons. Bake for 30 minutes longer. Yield: 10 servings.

Approx per serving: Cal 146, Prot 6.4 gr,
T Fat 8.7 gr, Chol 16.1 mg, Carbo 12.2 gr,
Sod 608.0 mg, Potas 240.0 mg.

Giselle McKenzie
Huntsville, AL

Microwave Vegetable Wreath

2 lb. broccoli, trimmed
Flowerets of 1 small head cauliflower
1 carrot, thinly sliced
1 zucchini, sliced ¼ in. thick
4 oz. mushrooms, sliced
1 sm. red pepper, cut into wide strips
¼ c. melted margarine

Arrange broccoli around outer edge of micro-proof platter with stalks toward center. Mound cauliflowerets in center. Arrange carrot, zucchini and mushrooms around cauliflower. Shape red pepper strips as for flower over cauliflower. Drizzle margarine over vegetables. Microwave, covered with plastic wrap, on High for 10 to 12 minutes or until vegetables are tender, turning platter every 4 minutes. Let stand for 5 minutes. Garnish with Parmesan cheese.
Yield: 8 servings.

Approx per serving: Cal 134, Prot 8.1 gr,
T Fat 6.4 gr, Chol 2.7 mg, Carbo 15.9 gr,
Sod 95.2 mg, Potas 946.9 mg.

Gina Watson
Ft. Wayne, IN

Microwave Vegetable Platter

Flowerets of 1 small bunch broccoli
Flowerets of 1 small head cauliflower
3 carrots, sliced
2 zucchini, sliced
¼ c. margarine, melted
¼ tsp. thyme
½ tsp. garlic salt
1 tomato, cut into wedges
6 oz. fresh mushroom halves
½ c. Parmesan cheese

Arrange broccoli, cauliflower, carrots and zucchini in desired pattern on microproof platter. Microwave, tightly covered, with plastic wrap for 6 minutes, turning 2 or 3 times. Drizzle mixture of margarine and seasonings over vegetables. Arrange tomato and mushrooms on platter. Sprinkle with cheese. Microwave, covered, for 1 to 2 minutes or until cheese melts. Yield: 6 servings.

Approx per serving: Cal 211, Prot 12.5 gr,
T Fat 11.0 gr, Chol 9.4 mg, Carbo 20.9 gr,
Sod 395.7 mg, Potas 1179.8 mg.

Jane Koonce
Golf, IL

Tangy Rice

4 c. long grain rice
Grated rind of 4 lemons
2 sticks butter
6 c. chicken stock
2 tsp. salt
¼ c. lemon juice
2 c. heavy cream

Sauté rice and lemon rind in butter in heavy saucepan for 5 minutes. Bring chicken stock to a boil in saucepan. Stir stock and salt into rice. Simmer, covered, for 20 minutes or until rice is tender and broth is absorbed. Stir in lemon juice. Add cream gradually, stirring constantly. Simmer over low heat for 5 minutes or until cream is absorbed. Add freshly ground pepper to taste. Yield: 12 servings.

Approx per serving: Cal 504, Prot 5.6 gr,
T Fat 30.6 gr, Chol 101.6 mg, Carbo 51.7 gr,
Sod 1038.1 mg, Potas 108.1 mg.

Jane Bigler
Shawnee Mission Pub. Sch., Lenexa, KS

Bombay Rice

½ med. onion, finely chopped
¼ tsp. cumin
1 tbsp. margarine
½ c. rice
1 tsp. tarragon vinegar
1½ c. boiling water

Sauté onion with cumin in margarine in saucepan until tender. Add rice. Sauté for several minutes. Add vinegar and water. Simmer, covered, for 20 minutes or until rice is tender. Spoon into serving dish. Yield: 2 servings.

Approx per serving: Cal 235, Prot 3.7 gr,
T Fat 6.0 gr, Chol 0.0 mg, Carbo 41.0 gr,
Sod 52.6 mg, Potas 111.7 mg.

Betsy Swindell
Houston, TX

Nutty Rice Patties

1½ c. cooked brown rice
½ c. shredded zucchini
1 egg, beaten
¼ c. whole wheat flour
1 tbsp. chopped parsley
1 tbsp. roasted sunflower seed
⅛ tsp. pepper
2 tbsp. oil
2 tbsp. brown sugar
2 tsp. cornstarch
2 tbsp. cider vinegar
⅓ c. pineapple juice
1 tbsp. orange juice

Combine first 7 ingredients in bowl; mix well. Shape into 6 patties. Brown rice patties on both sides in oil in skillet. Mix brown sugar, cornstarch, vinegar and juices in small saucepan. Cook over medium heat until thickened, stirring constantly. Serve sauce over patties. Yield: 6 servings.

Approx per serving: Cal 169, Prot 3.5 gr,
T Fat 6.6 gr, Chol 42.1 mg, Carbo 24.5 gr,
Sod 150.0 mg, Potas 145.5 mg.

Sharie Mueller
Jefferson County North H.S., Winchester, KS

Hominy Casserole

2 16-oz. cans yellow hominy, drained
1 4-oz. can chopped green chilies
1 2-oz. jar chopped pimento
2 c. grated Cheddar cheese
1 c. sour cream

Combine all ingredients in bowl; mix well. Spoon into buttered 2-quart casserole. Bake, covered, at 350 degrees for 45 minutes to 1 hour or until bubbly. Yield: 6 servings.

Approx per serving: Cal 320, Prot 12.8 gr,
T Fat 20.4 gr, Chol 54.2 mg, Carbo 21.4 gr,
Sod 622.2 mg, Potas 142.2 mg.

Patricia L. Sutton
Eldorado H.S., Albuquerque, NM

Whole Wheat Noodles

3 eggs, beaten
½ tsp. salt
1 c. all-purpose flour
1 c. whole wheat flour
6 c. chicken broth

Combine eggs, salt and flour in bowl; mix well. Roll ½ at a time very thin on floured surface. Let stand for 3 hours. Roll up loosely. Cut into strips. Cook in boiling chicken broth in large saucepan for 12 minutes or until tender. Yield: 6 servings.

Approx per serving: Cal 195, Prot 8.1 gr,
T Fat 3.4 gr, Chol 127.9 mg, Carbo 32.0 gr,
Sod 689.1 mg, Potas 0.74 mg.

Simone Hartfiel
College Station, TX

Stuffing Balls

½ c. chopped onion
½ c. chopped celery
2 tbsp. melted margarine
1 12-oz. can whole kernel corn, drained
1 8-oz. package seasoned stuffing mix
½ tsp. each pepper, marjoram
2 eggs, beaten
1 c. milk
½ c. melted margarine

Sauté onion and celery in 2 tablespoons margarine in saucepan until tender. Stir in corn, stuffing mix and seasonings. Add eggs and milk; mix well. Shape into balls. Place in 9x13-inch baking pan. Drizzle with ½ cup margarine. Bake at 375 degrees for 20 minutes or until brown. Yield: 8 servings.

Approx per serving: Cal 291, Prot 7.2 gr,
T Fat 16.2 gr, Chol 69.9 mg, Carbo 28.5 gr,
Sod 360.0 mg, Potas 181.2 mg.

Erica Montgomery
Stephens, OK

Honeyed Apple Rings

4 tart apples, cored, sliced
½ c. honey
2 tbsp. vinegar
¼ tsp. each salt, cinnamon

Arrange apples in foil-lined 9x13-inch baking pan. Drizzle with mixture of honey, vinegar, salt and cinnamon. Bake at 325 degrees for 10 minutes or until apples are tender. Yield: 8 servings.

Approx per serving: Cal 126, Prot 0.3 gr,
T Fat 0.0 gr, Chol 0.0 mg, Carbo 32.8 gr,
Sod 66.6 mg, Potas 115.0 mg.

Lori McKinney
Osage, OK

Low-Sodium Chutney

6 c. grated zucchini
2 green peppers, ground
2 c. grated tart apples
1 onion, ground
1 c. honey
12 oz. raisins
1 tbsp. celery seed
Juice and grated rind of 1 lemon
1⅓ c. vinegar
⅓ c. frozen orange juice concentrate

Combine all ingredients in stockpot; mix well. Simmer until thickened. Ladle into hot sterilized jars, leaving ½-inch headspace; seal with 2-piece lid. Process in boiling water bath for 10 minutes. Cool. Yield: 80 tablespoons.

Approx per tablespoon: Cal 34, Prot 0.4 gr,
T Fat 0.1 gr, Chol 0.0 mg, Carbo 9.0 gr,
Sod 2.1 mg, Potas 88.2 mg.

Sue Doerner
Spring Hill, TN

BREADS

Baking Powder Drop Biscuits

2 c. flour
2 tbsp. sugar
1 tbsp. baking powder
¼ tsp. salt
¼ c. shortening
1 c. milk

Sift dry ingredients into bowl. Cut in shortening until crumbly. Stir in milk. Drop by spoonfuls onto greased baking sheet. Bake at 450 degrees for 10 to 12 minutes or until light brown. Yield: 18 biscuits.
Note: Recipe may be used for strawberry shortcake.

Approx per biscuit: Cal 93, Prot 1.9 gr,
T Fat 3.7 gr, Chol 1.9 mg, Carbo 12.8 gr,
Sod 91.4 mg, Potas 33.5 mg.

Esther H. Nassar
Erie-Mason H.S., Toledo, OH

Microwave Banana Bread

2 tbsp. sugar
¼ c. chopped pecans
½ tsp. cinnamon
1½ c. flour
¾ c. sugar
½ tsp. salt
10 tbsp. melted margarine
⅓ c. milk
1 egg, beaten
2 med. bananas, mashed
¼ c. chopped pecans

Grease small glass bundt pan. Sprinkle with mixture of 2 tablespoons sugar, ¼ cup pecans and cinnamon. Combine remaining dry ingredients in bowl. Add margarine, milk, egg and bananas; mix well. Fold in remaining ¼ cup pecans. Spoon into prepared pan. Microwave on Medium for 9 minutes. Microwave on High for 3 to 5 minutes or until bread tests done. Let stand for several minutes. Remove to wire rack to cool. Yield: 12 slices.

Approx per slice: Cal 260, Prot 3.2 gr,
T Fat 14.0 gr, Chol 22.0 mg, Carbo 32.0 gr,
Sod 214.7 mg, Potas 136.1 mg.

Shari Rogers
Lincoln Mid. Sch., Abilene, TX

Banana Bread

½ c. margarine, softened
¾ c. sugar
2 eggs
1 c. all-purpose flour
1 tsp. soda
½ tsp. salt
1 c. whole wheat flour
3 lg. bananas, mashed
½ c. chopped walnuts
1 tsp. vanilla extract

Cream margarine and sugar in mixer bowl until light and fluffy. Add eggs 1 at a time, beating well after each addition. Sift all-purpose flour, soda and salt into bowl. Stir in whole wheat flour. Add to creamed mixture; mix well. Fold in bananas, walnuts and vanilla. Spoon into greased loaf pan. Bake at 350 degrees for 50 to 60 minutes or until bread tests done. Cool in pan for 10 minutes. Cool on wire rack. Yield: 12 slices.

Approx per slice: Cal 195, Prot 4.6 gr,
T Fat 5.0 gr, Chol 42.1 mg, Carbo 34.9 gr,
Sod 174.3 mg, Potas 190.8 mg.

Aileen Armijo Garcia
Sante Fe Tech. H.S., Sante Fe, NM

Bunker Hill Brown Bread

1½ c. flour
2 tsp. soda
1½ tsp. salt
1 c. wheat germ
1 c. graham cracker crumbs
2 eggs
⅓ c. oil
1 c. molasses
2 c. buttermilk

Sift flour, soda and salt into bowl. Mix in wheat germ and cracker crumbs. Add mixture of remaining ingredients; mix well. Spoon into 2 greased and floured tall 1-pound coffee cans. Do not cover. Bake at 375 degrees for 45 minutes or until bread tests done. Cool in cans for 10 minutes. Loosen with knife; turn out and slice. Serve hot. Yield: 24 slices.

Approx per slice: Cal 136, Prot 3.6 gr,
T Fat 4.5 gr, Chol 21.5 mg, Carbo 21.1 gr,
Sod 264.9 mg, Potas 221.4 mg.

Dorothy Moore
Central Jr. H.S., Sand Springs, OK

Raisin Brown Bread

2 c. whole wheat flour
1¼ c. raisins
1 c. buttermilk
½ c. molasses
2 tbsp. sugar
1 tsp. soda
¼ tsp. salt
2 tbsp. hot water
½ c. chopped pecans

Mix whole wheat flour, raisins, buttermilk, molasses and sugar in bowl. Dissolve soda and salt in hot water in small bowl. Add to flour mixture; mix well. Stir in pecans. Pour into greased and floured glass loaf pan. Bake at 350 degrees for 50 minutes. Yield: 12 slices.

Approx per slice: Cal 194, Prot 4.2 gr,
T Fat 4.0 gr, Chol 0.4 mg, Carbo 38.6 gr,
Sod 146.1 mg, Potas 372.9 mg.

Olive Ranck
Northeastern H.S., Williamsburg, IN

Poppy Seed Loaves

3 c. flour
2¼ c. sugar
1½ tsp. baking powder
1½ tbsp. poppy seed
½ tsp. salt
3 eggs
1 c. plus 2 tbsp. oil
1½ c. milk
1½ tsp. each almond, butter
  and vanilla extract
¾ c. confectioners' sugar
¼ c. orange juice
½ tsp. each almond, butter
  and vanilla extract

Combine first 5 ingredients in mixer bowl. Add eggs, oil, milk and 1½ teaspoons each flavoring; mix well. Spoon into waxed paper-lined loaf pans. Bake at 350 degrees for 1 hour. Remove to wire rack to cool; remove waxed paper. Frost with mixture of confectioners' sugar, orange juice and remaining flavorings. Yield: 24 slices.

Approx per slice: Cal 255, Prot 3.0 gr,
T Fat 11.6 gr, Chol 33.7 mg, Carbo 35.4 gr,
Sod 80.8 mg, Potas 51.0 mg.

Millie Morris
Richmond Hill H.S., Richmond Hill, GA

Pineapple-Carrot Bread

3 eggs
½ c. oil
½ c. melted butter
1½ c. sugar
2 tsp. vanilla extract
2 c. shredded carrots
1   8-oz. can crushed pineapple
  with liquid
3 c. flour
2 tsp. soda
½ tsp. baking powder
1½ tsp. cinnamon
¾ tsp. nutmeg
1 tsp. salt
½ c. raisins
1 c. chopped pecans

Beat eggs, oil, butter, sugar and vanilla in bowl until thick and smooth. Stir in carrots and pineapple. Add dry ingredients; mix just until moistened. Stir in raisins and pecans. Pour into 2 greased and floured 4½x8½-inch loaf pans. Bake at 325 degrees for 1 hour. Yield: 32 slices.

Approx per slice: Cal 183, Prot 2.4 gr,
T Fat 9.6 gr, Chol 32.6 mg, Carbo 22.8 gr,
Sod 168.1 mg, Potas 89.0 mg.

Kathleen Burchett
Area Supr. of H.E., State Dept. of Ed., Abingdon, VA

Golden Yam-Corn Bread

1 c. sifted flour
1 c. yellow cornmeal
4 tsp. baking powder
1 tsp. salt
¼ c. sugar
2 eggs
6 tbsp. milk
3 tbsp. oil
1   16-oz. can yams, mashed

Sift dry ingredients together. Combine eggs, milk and oil in bowl; mix well. Add yams; beat until blended. Stir in dry ingredients just until moistened. Spoon into greased 8-inch square baking pan. Bake at 425 degrees for 35 minutes or until bread tests done. Yield: 6 servings.

Photograph for this recipe on page 91.

Pumpkin Bread

4 eggs, beaten
3 c. sugar
1 c. oil
1 16-oz. can pumpkin
3½ c. flour
2 tsp. soda
1 tsp. baking powder
1 tsp. pumpkin pie spice
1 tsp. cinnamon
2 tsp. salt
1 c. mincemeat
⅔ c. water
1 c. chopped pecans

Beat eggs, sugar and oil in bowl until light. Add pumpkin. Add sifted dry ingredients; mix well. Add mincemeat, water and pecans. Beat until well mixed. Spoon into 4 well-oiled 1-pound coffee cans. Bake at 350 degrees for 1 hour. Cool bread in cans. Yield: 48 slices.
Note: Nutritional information does not include 1 cup mincemeat.

Approx per slice: Cal 149, Prot 1.8 gr,
T Fat 6.9 gr, Chol 21.0 mg, Carbo 20.6 gr,
Sod 159.4 mg, Potas 53.9 mg.

Annie Rust
Blanchard H.S., Norman, OK

Microwave Strawberry-Banana Bread

1 c. whole wheat flour
½ c. unbleached flour
⅔ c. packed brown sugar
1 tsp. soda
¼ tsp. nutmeg
½ c. reduced-calorie margarine, softened
⅓ c. buttermilk
2 eggs
1 c. sliced fresh strawberries
½ c. mashed banana
¼ c. chopped pecans

Line bottom of glass 5x9-inch loaf pan with waxed paper. Mix dry ingredients in mixer bowl. Add remaining ingredients. Beat at low speed until moistened. Beat at medium speed for 2 minutes, scraping bowl occasionally. Spoon into prepared loaf pan. Mold 2-inch foil strip over each end of pan, covering 1-inch of batter. Place on saucer in microwave oven. Microwave on Medium for 9 minutes, rotating dish ¼ turn every 3 minutes. Microwave on High for 3 minutes. Remove foil. Microwave for 2 to 9 minutes or until center springs back when lightly touched, turning dish once or twice. Let stand at room temperature for 5 to 10 minutes. Remove to wire rack to cool. Yield: 18 slices.
Note: Check manufacturer's instructions for your microwave oven regarding use of foil.
Nutritional information does not include reduced-calorie margarine.

Approx per slice: Cal 96, Prot 2.4 gr,
T Fat 2.0 gr, Chol 28.2 mg, Carbo 17.9 gr,
Sod 61.1 mg, Potas 116.2 mg.

Polly Anne Bryant
Lonoke Jr. H.S., Lonoke, AR

Zucchini Bread

⅔ c. shortening
2⅔ c. sugar
4 eggs
3 c. shredded zucchini
⅔ c. water
3⅓ c. flour
2 tsp. soda
½ tsp. baking powder
1 tsp. each cinnamon, cloves
1½ tsp. salt
2 tsp. vanilla extract
⅔ c. raisins

Grease bottom of three 4½x8½-inch loaf pans. Cream shortening and sugar in mixer bowl until light and fluffy. Add eggs, zucchini and water; mix well. Add dry ingredients gradually, mixing well after each addition. Stir in vanilla and raisins. Pour into prepared pans. Bake at 350 degrees for 60 to 70 minutes or until toothpick inserted in center comes out clean. Cool in pans for 5 minutes. Remove to wire rack to cool completely. Store, wrapped in plastic wrap, in refrigerator for up to 10 days. Yield: 30 slices.

Approx per slice: Cal 186, Prot 2.6 gr,
T Fat 5.9 gr, Chol 33.7 mg, Carbo 31.3 gr,
Sod 176.4 mg, Potas 73.4 mg.

Fran Heckman
Waupaca H.S., Waupaca, WI

Chocolate-Pecan-Zucchini Bread

3 eggs, slightly beaten
2 c. sugar
1 c. oil
2 tsp. vanilla extract
2 c. grated zucchini
3 c. flour
2 tbsp. cocoa
1 tsp. soda
½ tsp. baking powder
1 tsp. each cinnamon, salt
1 c. chopped pecans

Beat eggs, sugar, oil and vanilla in bowl. Add zucchini; mix well. Sift in dry ingredients; stir just until moistened. Fold in pecans. Spoon into 2 greased and floured loaf pans. Bake at 325 degrees for 50 minutes or until bread tests done. Remove to wire rack to cool. Let stand, wrapped, for 24 hours before slicing. Yield: 24 slices.

Approx per slice: Cal 248, Prot 3.1 gr,
T Fat 13.5 gr, Chol 31.6 mg, Carbo 29.9 gr,
Sod 138.1 mg, Potas 81.9 mg.

Edna O. Hutchens
Heritage Sch., Maryville, TN

Honey-Zucchini Bread

2 c. all-purpose flour
1 c. whole wheat flour
1 tsp. soda
½ tsp. baking powder
1 tsp. salt
1 tbsp. cinnamon
3 eggs, slightly beaten
⅔ c. oil
1⅔ c. honey
1 tbsp. vanilla extract
2 c. grated zucchini
1 c. finely chopped pecans

Mix dry ingredients in large bowl. Add mixture of eggs, oil, honey, vanilla and zucchini; mix just until moistened. Fold in pecans. Spoon into waxed paper-lined loaf pans. Bake at 325 degrees for 1 hour. Cool in pan for 10 minutes. Remove; cool on wire rack. Yield: 20 slices.

Approx per slice: Cal 270, Prot 3.9 gr,
T Fat 12.6 gr, Chol 37.9 mg, Carbo 38.2 gr,
Sod 167.0 mg, Potas 120.0 mg.

Ann Hodgson
Dodgeland H.S., Juneau, WI

Banana Muffins

1 egg
⅓ c. oil
½ c. sugar
1 c. mashed banana
1½ c. buttermilk baking mix

Beat egg in bowl. Blend in oil and sugar. Add banana; mix well. Add baking mix; stir just until moistened. Fill paper-lined muffin cups ⅔ full. Bake at 375 degrees for 15 to 18 minutes or until muffins test done. Yield: 14 muffins.

Approx per muffin: Cal 147, Prot 1.6 gr,
T Fat 7.2 gr, Chol 18.1 mg, Carbo 19.5 gr,
Sod 171.7 mg, Potas 74.6 mg.

Shirley Henkel
Statesville H.S., Statesville, NC

Bostonian Muffins

1½ c. flour
½ c. whole wheat flour
1 c. sugar
1 tbsp. baking powder
½ tsp. salt
½ tsp. cinnamon
1½ to 2 c. blueberries
¼ c. melted butter
½ c. milk
2 eggs, beaten
½ tsp. vanilla extract
1½ tbsp. sugar

Combine flours, 1 cup sugar, baking powder, salt and cinnamon in bowl. Toss 1 tablespoon flour mixture with blueberries in bowl. Mix butter, milk, eggs and vanilla in bowl. Add remaining flour mixture; mix just until moistened. Stir in blueberries. Spoon into buttered muffin cups. Sprinkle with 1½ tablespoons sugar. Bake at 425 degrees for 15 minutes. Cool in pan for 5 minutes. Serve warm. Yield: 12 servings.

Approx per serving: Cal 214, Prot 3.9 gr,
T Fat 5.5 gr, Chol 55.4 mg, Carbo 38.0 gr,
Sod 233.8 mg, Potas 81.1 mg.

Amy Eilert
Los Alamos, NM

Six-Week Bran Muffins

1 c. 100% bran cereal
½ c. shortening
1 c. boiling water
2 eggs, beaten
½ c. sugar
2 c. buttermilk
2½ c. flour
2½ tsp. soda
1 tsp. salt
2 c. 100% bran cereal

Place 1 cup bran cereal and shortening in bowl. Add boiling water; mix well. Let stand for several minutes. Beat eggs, sugar and buttermilk in bowl until smooth. Add flour, soda and salt; mix well. Stir in bran mixture and remaining 2 cups cereal. Store, covered, in refrigerator for up to 6 weeks. Fill greased muffin cups ⅔ full. Bake at 400 degrees for 20 minutes. Yield: 36 muffins.

Approx per muffin: Cal 94, Prot 2.4 gr,
T Fat 3.6 gr, Chol 14.3 mg, Carbo 15.0 gr,
Sod 168.1 mg, Potas 60.5 mg.

Nora Sweat
West Hardin H.S., Elizabethtown, KY

Six-Week Muffins

5 c. flour
5 tsp. soda
2 tsp. salt
1   15-oz. package raisin bran cereal
3 c. sugar
4 eggs
1 qt. buttermilk
1 c. oil

Sift flour, soda and salt into bowl. Mix in cereal and sugar. Combine eggs, buttermilk and oil in bowl; mix well. Add to dry ingredients; mix well. Store in covered container in refrigerator for up to 6 weeks. Fill greased muffin cups ⅔ full. Bake at 400 degrees for 15 minutes. Yield: 36 muffins.

Approx per muffin: Cal 221, Prot 4.4 gr,
T Fat 7.1 gr, Chol 28.6 mg, Carbo 40.0 gr,
Sod 322.2 mg, Potas 96.4 mg.

Janice Sapp
Claxton H.S., Claxton, GA
Pat Vaughan
Fairfield H.S., Fairfield, IL

Quickie Bran Muffins

1 c. low-fat milk
1 egg, beaten
¼ c. oil
1 c. bran cereal
1 c. whole wheat flour
¼ c. sugar
1 tbsp. baking powder
½ tsp. salt

Combine milk, egg and oil in bowl. Mix in cereal. Sift in remaining ingredients; mix well. Spoon into muffin cups sprayed with nonstick cooking spray. Bake at 400 degrees for 20 minutes. Serve hot with honey and butter.
Yield: 12 muffins.

Approx per muffin: Cal 120, Prot 3.2 gr,
T Fat 5.6 gr, Chol 21.9 mg, Carbo 17.3 gr,
Sod 162.4 mg, Potas 104.2 mg.

Dorothy Moore
Central Jr. H.S., Sand Springs, OK

Oatmeal Muffins

1 c. whole wheat flour
¾ c. quick-cooking oats
¼ c. packed brown sugar
1 tbsp. baking powder
½ tsp. each soda, salt and cinnamon
½ c. raisins
1 egg
1 c. milk
¼ c. oil

Mix flour, oats, brown sugar, baking powder, soda, salt, cinnamon and raisins in bowl. Blend egg, milk and oil in small bowl. Add to dry ingredients; stir just until moistened. Spoon into greased muffin cups. Bake at 400 degrees for 20 to 25 minutes or until brown. Yield: 12 muffins.

Approx per muffin: Cal 148, Prot 3.4 gr,
T Fat 6.3 gr, Chol 23.9 mg, Carbo 20.9 gr,
Sod 223.8 mg, Potas 152.3 mg.

Barbara Tripp
Spring Lake Jr.-Sr. H.S., Spring Lake, MI

Oatmeal-Raisin Muffins

1 c. milk
1 tsp. vinegar
1 egg, beaten
⅓ c. packed brown sugar
⅓ c. oil
½ to 1 c. raisins
1 c. oats
1 c. flour
1 tsp. baking powder
½ tsp. soda

Mix milk and vinegar in bowl. Let stand for 5 minutes. Add egg, brown sugar, oil and raisins; mix well. Add mixture of remaining dry ingredients; stir just until moistened. Fill greased muffin cups ⅔ full. Bake at 400 degrees for 15 to 20 minutes or until brown. Yield: 12 muffins.

Approx per muffin: Cal 195, Prot 3.6 gr,
T Fat 7.9 gr, Chol 23.9 mg, Carbo 28.8 gr,
Sod 82.3 mg, Potas 181.8 mg.

Janie Armstrong
Chippewa Valley H.S., Royal Oak, MI

Orange-Blueberry Muffins

1   13-oz. package blueberry muffin mix
1 egg, slightly beaten
⅓ c. Florida frozen orange juice
*   concentrate, thawed*
2 tbsp. plus 2 tsp. water

Combine all ingredients in bowl; mix well. Spoon into 12 greased muffin cups. Bake at 400 degrees for 15 to 20 minutes or until brown. Cool in pan for 5 minutes. Yield: 12 muffins.

Photograph for this recipe on page 9.

Nutty Whole Grain Muffins

⅓ c. peanut butter
⅔ c. packed brown sugar
1 egg
1 c. milk
1 c. flour
1 c. quick-cooking oats
½ c. wheat germ
1 tbsp. baking powder
½ tsp. each cinnamon, salt

Cream peanut butter, brown sugar and egg in mixer bowl until light. Blend in milk. Add mixture of dry ingredients; mix just until moistened. Fill greased medium muffin cups ⅔ full. Bake at 400 degrees for 18 to 20 minutes or until toothpick inserted in center comes out clean. Serve warm. Yield: 12 muffins.

Approx per muffin: Cal 174, Prot 5.2 gr,
T Fat 5.4 gr, Chol 23.9 mg, Carbo 27.0 gr,
Sod 233.3 mg, Potas 158.3 mg.

Sandra Whaley
North Whitfield Mid. Sch., Dalton, GA

Zucchini and Carrot Muffins

1⅓ c. all-purpose flour
⅓ c. whole wheat flour
⅔ c. packed light brown sugar
½ tsp. each soda, salt
1 tsp. cinnamon
¼ tsp. nutmeg
⅓ c. safflower oil
¼ c. skim milk
3 egg whites
1¼ c. grated carrots
⅔ c. grated zucchini
½ c. chopped walnuts

Mix dry ingredients in bowl. Combine oil, milk, and egg whites in large bowl; mix well. Add dry ingredients; mix just until moistened. Stir in carrots, zucchini and walnuts. Fill oiled muffin cups ⅔ full. Bake at 425 degrees for 20 minutes or until toothpick inserted in center comes out clean. Serve warm. Yield: 12 muffins.

Approx per muffin: Cal 205, Prot 3.9 gr,
T Fat 9.5 gr, Chol 0.1 mg, Carbo 27.2 gr,
Sod 147.3 mg, Potas 162.7 mg.

Tranetta L. McComb
Barrington H.S., Barrington, IL

Whole Wheat Buttermilk Pancakes

1 c. buttermilk
2 tbsp. oil
1 egg
½ c. whole wheat flour
½ c. all-purpose flour
1 tbsp. sugar
1 tsp. baking powder
½ tsp. soda
½ tsp. salt

Beat buttermilk, oil and egg in mixer bowl until smooth. Add mixture of dry ingredients; mix just until moistened. Spoon onto hot lightly greased griddle. Bake until light brown on both sides. Yield: 10 pancakes.
Note: May substitute mixture of wheat germ and all-purpose flour for ½ cup all-purpose flour.

Approx per pancake: Cal 89, Prot 3.0 gr, T Fat 3.5 gr, Chol 25.8 mg, Carbo 11.6 gr, Sod 218.8 mg, Potas 69.4 mg.

Beverly Nickel
Newton H.S., Newton, KS

Cheese Knots

⅔ c. crushed rice Chex cereal
3 tbsp. Parmesan cheese
2 tbsp. melted margarine
1 10-count can refrigerator biscuits

Combine cereal, cheese and margarine in bowl; mix well. Separate biscuits; cut into halves. Coat each biscuit half with crumb mixture. Roll each biscuit into 8-inch roll, pressing crumbs into roll. Shape into knot on ungreased baking sheet. Bake at 425 degrees for 6 to 8 minutes or until light brown. Serve immediately as snack or accompaniment for quick lunch or supper. Yield: 20 knots.
Note: May substitute ⅔ cup crushed corn Chex cereal or ½ cup crushed wheat Chex cereal for rice Chex cereal.

Approx per knot: Cal 68, Prot 1.8 gr, T Fat 2.6 gr, Chol 8.8 mg, Carbo 9.2 gr, Sod 176.3 mg, Potas 14.4 mg.

Lavern Frentzel
Perry Co. Dist. 32 Sch., Uniontown, MO

Soft Pretzels

2 c. all-purpose flour
1 c. whole wheat flour
1 pkg. fast-rising dry yeast
¾ tsp. salt
1¼ c. water
2 tbsp. honey
1 c. all-purpose flour
1 egg
1 tbsp. water
2 tbsp. sesame seed

Combine 2 cups all-purpose flour, whole wheat flour, yeast and salt in mixer bowl. Heat water and honey in saucepan to 125 to 130 degrees. Add to dry ingredients; mix well. Add enough remaining 1 cup all-purpose flour to make soft dough. Knead on floured surface for 5 to 7 minutes or until smooth and elastic. Divide into 12 portions. Roll each portion into rope on floured surface; shape into pretzel. Place on greased baking sheets. Brush with mixture of egg and 1 tablespoon water. Sprinkle with sesame seed. Bake at 425 degrees for 15 to 20 minutes or until brown. Serve with mustard. Yield: 12 pretzels.

Approx per pretzel: Cal 174, Prot 5.6 gr, T Fat 1.7 gr, Chol 21.1 mg, Carbo 34.3 gr, Sod 140.5 mg, Potas 95.2 mg.

Barbara Bird
Alma H.S., Alma, MI

Bran Yeast Rolls

1½ c. water
2 c. shreds of wheat bran cereal
1 c. butter
2 tsp. salt
¾ c. sugar
2 pkg. dry yeast
½ c. 110-degree water
2 eggs, beaten
5 c. (about) flour

Bring 1½ cups water to a boil in saucepan. Combine with cereal, butter, salt and sugar in mixer bowl. Cool to 105 to 115 degrees. Dissolve yeast in ½ cup 110-degree water in large mixer bowl. Let stand for 5 minutes. Add cereal mixture, eggs and 2 cups flour. Beat at medium speed for 1 minute. Add remaining flour gradually, mixing well after each addition. Place

in greased bowl, turning to grease surface. Chill, covered, for 2 hours. Punch dough down on floured surface. Shape into 2-inch balls. Place in greased muffin cups. Let rise, covered, in warm place for 1 hour or until doubled in bulk. Bake at 375 degrees for 20 minutes. Yield: 36 rolls.
Note: Nutritional information does not include wheat bran cereal.

Approx per roll: Cal 130, Prot 2.4 gr,
T Fat 5.6 gr, Chol 29.8 mg, Carbo 17.6 gr,
Sod 184.7 mg, Potas 29.4 mg.

Ruth Irwin
Shawnee H.S., Cape Girardeau, MO

Special Dinner Rolls

1 pkg. dry yeast
1 tbsp. sugar
¼ c. warm water
1 c. milk, scalded, cooled
¾ tsp. salt
½ c. melted butter
3 eggs, well beaten
½ c. sugar
3½ to 4 c. flour

Dissolve yeast and 1 tablespoon sugar in warm water in large bowl. Add milk, salt, butter, eggs and ½ cup sugar; mix well. Add enough flour to make soft dough. Place in greased bowl, turning to grease surface. Let rise in warm place for 5 to 6 hours or in refrigerator overnight. Shape as desired; place in greased baking pan. Let rise until doubled in bulk. Bake at 325 degrees for 20 minutes or until golden brown. Yield: 18 rolls.

Approx per roll: Cal 194, Prot 4.7 gr,
T Fat 6.8 gr, Chol 59.8 mg, Carbo 28.2 gr,
Sod 168.9 mg, Potas 66.1 mg.

Kay Caskey
Swope Mid. Sch., Reno, NV

Carrot-Raisin Bread

1 pkg. dry yeast
½ c. 110 to 120-degree water
1 c. flour
¼ c. sugar
1 egg
¼ c. shortening
2 tbsp. dry milk powder
½ c. warm water

½ c. raisins
3 c. flour
1 c. grated carrots
1 tsp. cinnamon
¼ tsp. allspice
⅛ tsp. cloves

Dissolve yeast in ½ cup warm water in mixer bowl. Add next 6 ingredients. Beat for 2 minutes. Mix raisins with remaining 3 cups flour. Add raisin mixture, carrots and spices to batter; mix well. Place in greased bowl, turning to grease surface. Let rise, covered, for 20 minutes. Shape into 2 loaves. Place in 5x9-inch baking pans. Let rise, covered, for 45 minutes or until doubled in bulk. Bake at 375 degrees for 30 minutes or until golden. Yield: 24 slices.

Approx per slice: Cal 121, Prot 2.8 gr,
T Fat 2.8 gr, Chol 10.6 mg, Carbo 21.0 gr,
Sod 8.0 mg, Potas 73.2 mg.

Debra Hart
Emerson Jr. H.S., Enid, OK

High Protein Bread

1½ c. cottage cheese
1 c. buttermilk
3 pkg. dry yeast
⅓ c. sugar
4 tsp. salt
¾ tsp. soda
3 eggs
7½ c. sifted unbleached flour
¼ c. safflower oil

Mix cottage cheese and buttermilk in saucepan. Heat to 110 to 120 degrees. Mix in yeast. Let stand for several minutes. Stir in sugar, salt and soda. Combine with eggs and 2 cups flour in bowl; mix well. Add remaining flour gradually, mixing well after each addition. Mix in oil. Place in greased bowl, turning to grease surface. Let rise, covered, in warm place until doubled in bulk. Knead on floured surface for 8 to 10 minutes or until smooth and elastic. Shape into 3 loaves; place each in greased 5x9-inch loaf pan. Let rise until doubled in bulk. Bake at 400 degrees for 30 minutes. Yield: 30 slices.

Approx per slice: Cal 164, Prot 6.1 gr,
T Fat 3.2 gr, Chol 27.8 mg, Carbo 27.1 gr,
Sod 350.5 mg, Potas 72.1 mg.

Olive Ranck
Northeastern H.S., Williamsburg, IN

Three-Grain Bread

1 pkg. dry yeast
3 tbsp. honey
3 c. warm water
2 c. whole wheat flour
4½ c. all-purpose flour
¼ c. oats
½ c. wheat germ
¾ c. bran
¾ c. nonfat dry milk powder
1½ tsp. salt
2 eggs, beaten
½ c. margarine, softened

Dissolve yeast and honey in warm water in large bowl. Let stand for several minutes. Add mixture of dry ingredients, eggs and margarine; mix well. Let stand for 5 minutes. Knead on floured surface for 5 minutes or until smooth and elastic. Shape into 3 loaves. Place on greased baking sheet. Let stand in warm place for 10 minutes. Bake at 350 degrees for 30 to 40 minutes or until brown. Cool on wire rack. Yield: 30 slices.

Approx per slice: Cal 154, Prot 5.0 gr,
T Fat 4.1 gr, Chol 17.2 mg, Carbo 25.4 gr,
Sod 167.1 mg, Potas 114.0 mg.

Marian E. Baker
Sycamore H.S., Sycamore, IL

Crusty Honey Whole Wheat Bread

2 c. all-purpose flour
1 c. whole wheat flour
2 pkg. dry yeast
1 tbsp. salt
1 c. milk
1 c. water
½ c. honey
3 tbsp. margarine
1 egg
1½ to 2 c. all-purpose flour
1½ c. whole wheat flour

Combine 2 cups all-purpose flour, 1 cup whole wheat flour, yeast and salt in large mixer bowl. Heat milk, water, honey and margarine to 120 degrees in saucepan. Add warm milk mixture and egg to dry ingredients. Beat with dough hook for 3 minutes or by hand until smooth. Add enough remaining flours gradually to make stiff dough. Knead with dough hook for 4 minutes or by hand for 5 to 8 minutes or until smooth and elastic. Place in greased bowl, turning to grease surface. Let rise, covered, in warm place for 1 hour or until doubled in bulk. Shape into 2 loaves. Place in greased 5x9-inch pans. Let rise, covered, for 1½ hours or until doubled in bulk. Bake at 375 degrees for 10 minutes. Reduce temperature to 350 degrees. Bake for 30 to 35 minutes longer or until brown. Cool on wire rack. Yield: 24 slices.

Approx per slice: Cal 163, Prot 4.7 gr,
T Fat 2.5 gr, Chol 12.0 mg, Carbo 31.2 gr,
Sod 293.1 mg, Potas 99.0 mg.

Loretta Briggs
Breckenridge Comm. Sch., Wheeler, MI

High Protein Wheat Bread

2 c. bread flour
2 tsp. salt
2 pkg. dry yeast
1 c. water
½ c. honey
¼ c. margarine
1 c. cream-style cottage cheese
2 eggs
1 c. whole wheat flour
½ c. oats
1 c. chopped pecans
2 to 3 c. bread flour

Mix 2 cups bread flour, salt and yeast in mixer bowl. Heat water, honey, margarine and cottage cheese in saucepan to 120 to 130 degrees. Add warm mixture and eggs to flour mixture; blend well. Beat at high speed for 3 minutes. Stir in whole wheat flour, oats, pecans and enough remaining bread flour to make soft dough. Knead on floured surface for 10 minutes or until smooth and elastic. Place in greased bowl, turning to grease surface. Let rise, covered, in warm place for 1 hour or until doubled in bulk. Punch dough down. Let rest, covered, for 15 minutes. Shape into 2 loaves. Place in greased 5x9-inch loaf pans. Let rise until doubled in bulk. Bake at 375 degrees for 35 to 40 minutes or until brown. Remove to wire rack to cool. Yield: 24 slices.

Approx per slice: Cal 209, Prot 6.3 gr,
T Fat 6.8 gr, Chol 23.0 mg, Carbo 31.6 gr,
Sod 230.8 mg, Potas 108.6 mg.

Trudy K. Miller
Bishop Carroll H.S., Wichita, KS

DESSERTS

Apple Burritos

6 lg. tart apples, peeled, sliced
½ c. sugar
2 tbsp. cornstarch
2 tbsp. dry white wine
2 tsp. cinnamon
¼ tsp. allspice
8 lg. flour tortillas
6 tbsp. butter

Combine apples and sugar with water to cover in saucepan. Simmer for 5 minutes or until tender. Blend cornstarch, wine and spices in small bowl. Add several drops of water to dissolve cornstarch if necessary. Stir into apples. Cook until thick, stirring constantly. Soften tortillas in 2 tablespoons butter in skillet for 10 to 15 seconds, turning once. Add remaining butter as necessary. Spoon scant ¼ cup apple filling onto each tortilla; roll to enclose filling. Place seam side down on oiled rack in broiler pan. Brush with remaining butter. Broil for 4 to 5 minutes or until lightly browned. Serve with unflavored yogurt or sour cream. Yield: 8 servings.
Note: May sprinkle filled tortillas with cinnamon-sugar before broiling if desired.

Approx per serving: Cal 303, Prot 2.9 gr,
T Fat 10.5 gr, Chol 27.5 mg, Carbo 51.7 gr,
Sod 250.7 mg, Potas 210.1 mg.

Suva Bastin
Clarksville, TN

Carrot Pudding

4 eggs
½ c. sugar
½ tsp. salt
½ c. grated carrots
1 tsp. vanilla extract
1 qt. milk

Beat eggs, sugar and salt in mixer bowl until thick and lemon-colored. Add carrots and vanilla; mix well. Mix in milk. Pour into small glass baking dish. Place in pan of water. Bake at 300 degrees for 1 hour. Yield: 2 servings.

Approx per serving: Cal 685, Prot 30.3 gr,
T Fat 28.7 gr, Chol 573.9 mg, Carbo 77.2 gr,
Sod 912.7 mg, Potas 927.5 mg.

Kathleen Burchett
Area Supr. of H.E., State Dept. of Ed., Abingdon, VA

Citrus Cheesecake

⅔ c. cornflakes
¼ c. Grape Nuts
2 tsp. melted butter
1 tsp. brown sugar
1 tsp. grated orange rind
1 tsp. grated lemon rind
2 envelopes unflavored gelatin
¼ c. sugar
1½ c. low-fat milk
3 eggs, separated
3 c. ricotta cheese
1 tbsp. orange juice
1 tbsp. lemon juice
1 tsp. grated orange rind
1 tsp. grated lemon rind
1 tsp. vanilla extract
½ c. sugar
2 kiwifruit, peeled, sliced

Process cornflakes and Grape Nuts in food processor until finely crushed. Combine with next 4 ingredients in bowl; mix well. Press over bottom of 8-inch springform pan. Bake at 350 degrees for 5 minutes. Cool. Mix gelatin and ¼ cup sugar in double boiler. Beat in milk and egg yolks. Let stand until gelatin softens. Cook over boiling water until gelatin dissolves and mixture coats spoon, stirring constantly. Chill until partially set. Process ricotta cheese in food processor until smooth. Combine with next 5 ingredients in bowl; mix well. Stir in partially congealed mixture. Beat egg whites until frothy. Add ½ cup sugar gradually, beating constantly until soft peaks form. Fold gently into ricotta cheese mixture. Spoon into prepared pan. Chill for 8 hours or longer. Place on serving plate; remove side of pan. Arrange kiwifruit on top. Yield: 12 servings.
Note: Nutritional information does not include kiwifruit or cholesterol value for ricotta cheese.

Approx per serving: Cal 188, Prot 10.0 gr,
T Fat 8.4 gr, Chol 66.4 mg, Carbo 19.9 gr,
Sod 124.8 mg, Potas 78.3 mg.

Nicole Schmidt
Tacoma, WA

Healthy Dessert Casserole

2 c. flour
2 c. sugar
1 tsp. soda
1 tsp. cinnamon

½ tsp. cloves
¼ tsp. salt
⅔ c. shortening
2 c. chopped apple
2 c. grated carrots
1 c. raisins
½ c. chopped pecans

Sift dry ingredients together into bowl. Cut in shortening until crumbly. Add apple, carrots, raisins and pecans; mix well. Spoon into 8x12-inch baking dish. Bake at 350 degrees for 30 minutes. Serve warm with whipped topping. Yield: 8 servings.

Approx per serving: Cal 604, Prot 4.8 gr,
T Fat 24.5 gr, Chol 0.0 mg, Carbo 95.8 gr,
Sod 188.5 mg, Potas 342.1 mg.

Linda Finley
Harrison Central Ninth Gr. Sch., Gulfport, MS

Charles' Fruitsicle

1 6-oz. can frozen grape juice
   concentrate
2 6-oz. cans water
4 wooden sticks

Combine grape juice and water in pitcher; mix well. Pour into popsicle molds or small paper cups. Freeze for 1 hour. Place wooden stick in center of each fruitsicle. Freeze until firm. Yield: 4 servings.

Approx per serving: Cal 99, Prot 0.3 gr,
T Fat 0.0 gr, Chol 0.0 mg, Carbo 25.0 gr,
Sod 1.5 mg, Potas 63.8 mg.

Danna Sue Hadsock
Appling Co. Jr. H.S., Baxley, GA

Special Occasion Glazed Fruit

1 16-oz. can pineapple chunks
1 11-oz. can mandarin oranges
1 sm. package vanilla instant pudding mix
¼ c. orange instant breakfast drink mix
1 20-oz. can cherry pie filling
2 tbsp. poppy seed
2 to 3 bananas, sliced

Drain pineapple and oranges, reserving juices. Combine reserved juices, instant pudding mix and instant breakfast drink mix in bowl; mix well. Add pineapple, oranges, pie filling and poppy seed. Chill overnight. Stir in bananas gently just before serving. Yield: 8 servings.

Approx per serving: Cal 243, Prot 1.5 gr,
T Fat 0.5 gr, Chol 0.0 mg, Carbo 61.6 gr,
Sod 66.4 mg, Potas 282.0 mg.

Patricia B. Foster
Scott Mid. Sch., Scott, LA

Fruit Dip

1 c. marshmallow creme
8 oz. cream cheese, softened
1 tsp. orange extract
¼ tsp. nutmeg

Combine all ingredients in bowl; mix well. Chill in refrigerator. Serve with bite-sized fresh fruit on skewers. Yield: 32 tablespoons.

Approx per tablespoon: Cal 46, Prot 0.6 gr,
T Fat 2.7 gr, Chol 7.9 mg, Carbo 5.2 gr,
Sod 21.4 mg, Potas 7.3 mg.

Billie L. Perrin
Lafayette Co. C-1 H.S., Higginsville, MO

Winter Fruit Soup

1 c. each seedless raisins, pitted prunes
3 c. sliced apples
2 c. sliced peaches
1 orange, sliced
1 tbsp. lemon juice
¼ tsp. each cloves, nutmeg
1 tsp. cinnamon
2 c. boiling water
¼ c. cornstarch
2 c. pineapple juice
¼ c. (or less) sugar

Combine first 10 ingredients in large saucepan. Cook, covered, until fruit is tender. Mix cornstarch with a small amount of water. Add enough hot water to measure 2 cups. Stir into fruit mixture. Cook over medium heat until slightly thickened. Stir in pineapple juice and sugar to taste. Serve hot or cold garnished with a slice of lime. Yield: 16 servings.

Approx per serving: Cal 134, Prot 0.9 gr,
T Fat 0.3 gr, Chol 0.0 mg, Carbo 34.6 gr,
Sod 4.7 mg, Potas 279.0 mg.

Connie Layman
Celina, TX

Grapefruit Hot Fudge Sundae

6 oz. milk chocolate, broken
½ c. sweetened condensed milk
1 to 2 tbsp. milk
2 lg. Florida grapefruit
¼ c. packed brown sugar
4 scoops vanilla ice cream

Place chocolate and sweetened condensed milk in top of double boiler. Cook over hot water until chocolate is melted, stirring until smooth. Add milk to thin to desired consistency; keep warm. Cut grapefruit into halves; remove seed. Loosen sections from membrane with serrated knife. Place on baking sheet. Sprinkle with brown sugar. Broil grapefruit 3 to 5 inches from heat source for 5 minutes or until golden and bubbly. Place in serving bowls. Top with scoop of ice cream. Spoon chocolate sauce over top. Yield: 4 servings.

Photograph for this recipe on page 9.

Judy's Granola

5 c. oats
1 c. chopped dates
1 c. raisins
½ c. sunflower seed
1 c. wheat germ
2 c. flour
1 c. instant nonfat dry milk powder
⅔ c. oil
¾ c. honey
2 tsp. vanilla extract

Combine first 7 ingredients in bowl. Add mixture of oil, honey and vanilla; mix well. Spread on shallow baking sheet. Sprinkle with salt to taste. Bake at 300 degrees for 30 minutes, stirring every 10 minutes. Cool. Store in airtight container. Yield: 9 cups.

Approx per cup: Cal 679, Prot 19.0 gr,
T Fat 28.8 gr, Chol 15.5 mg, Carbo 91.6 gr,
Sod 66.3 mg, Potas 798.8 mg.

Judy Murray
Lake Mills H.S., Lake Mills, WI

Vickie's Granola

2 c. oats
1 c. coconut
1½ c. chopped walnuts

½ c. wheat germ
1 tsp. salt
¼ c. oil
1 recipe Sweetened Condensed Milk
  Substitute
1½ c. raisins

Combine first 5 ingredients in bowl; mix well. Stir in oil and Sweetened Condensed Milk Substitute. Place on waxed paper-lined baking sheet. Bake at 300 degrees for 1 hour, stirring every 15 minutes. Cool in pan, stirring occasionally. Add raisins. Store in airtight container. Yield: 6 cups.

Sweetened Condensed Milk Substitute

1 c. instant nonfat dry milk powder
⅔ c. packed brown sugar
3 tbsp. melted butter
⅓ c. boiling water

Combine all ingredients in blender container. Process until smooth. Use in place of 14-ounce can sweetened condensed milk. Yield: 1 recipe.

Approx per cup: Cal 871, Prot 17.7 gr,
T Fat 47.1 gr, Chol 41.0 mg, Carbo 103.7 gr,
Sod 563.2 mg, Potas 1037.5 mg.

Vickie J. Hadley
Paul Harding H.S., Fort Wayne, IN

Ice Cream Dessert

2 c. flour
½ c. oats
½ c. packed brown sugar
1½ c. chopped pecans
2 sticks margarine, melted
2 qt. vanilla ice cream, softened
1 12-oz. jar chocolate ice cream topping

Line baking sheet with foil. Combine flour, oats, brown sugar, pecans and margarine in bowl; mix well. Spread on prepared baking sheet. Bake at 300 degrees for 30 to 35 minutes or until light brown, stirring occasionally. Sprinkle half the mixture in greased 3-quart dish. Layer ice cream and topping over crumbs. Top with remaining crumbs. Freeze for several hours to 1 week. Yield: 12 servings.

Approx per serving: Cal 705, Prot 8.6 gr,
T Fat 47.4 gr, Chol 56.2 mg, Carbo 67.4 gr,
Sod 256.3 mg, Potas 357.4 mg.

Linda Wright
Idabel H.S., Foreman, AR

Natillas (Soft Custard)

4 eggs, separated
¼ c. flour
4 c. milk
¾ c. sugar
⅛ tsp. salt

Beat egg yolks, flour and 1 cup milk in mixer bowl until smooth. Combine remaining 3 cups milk, sugar and salt in saucepan. Bring just to the boiling point. Stir in egg yolk mixture gradually. Cook until thickened, stirring constantly. Cool to room temperature. Beat egg whites in bowl until stiff but not dry. Fold gently into custard. Spoon into serving dish. Chill until serving time. Garnish with nutmeg and cinnamon. Yield: 6 servings.

Approx per serving: Cal 275, Prot 10.6 gr,
T Fat 9.6 gr, Chol 191.3 mg, Carbo 37.1 gr,
Sod 166.9 mg, Potas 283.1 mg.

Aileen Armijo Garcia
Santa Fe Tech. H.S., Santa Fe, NM

Bronzed Pears on a Spiced Cloud

1 egg
2 tbsp. brown sugar
½ tsp. vanilla extract
⅛ tsp. allspice
1 c. whipped topping
1 tbsp. butter
1 tbsp. brown sugar
2 Western winter pears, cut into halves

Combine egg, 2 tablespoons brown sugar, vanilla and allspice in mixer bowl. Beat at high speed until smooth and thick. Fold in whipped topping. Chill, covered, until serving time. Melt butter in large skillet. Blend in 1 tablespoon brown sugar. Place pears cut side down in skillet. Cook, covered, over medium heat for 5 minutes. Turn pears. Cook, covered, for 5 minutes longer. Spoon chilled mixture into dessert dishes. Place pear half in each. Spoon brown sugar mixture over pears. Garnish with nutmeg. Yield: 4 servings.

Photograph for this recipe on page 101.

Frozen Strawberry Yogurt

1 env. unflavored gelatin
½ c. milk
2 tbsp. sugar
1 tbsp. lemon juice
2 tbsp. light corn syrup
2 eggs, beaten
2 c. yogurt
½ c. strawberries
⅛ tsp. salt
2 tsp. vanilla extract

Soften gelatin in milk in saucepan. Heat over low heat until dissolved, stirring constantly; cool slightly. Beat sugar, lemon juice and corn syrup gradually into eggs. Add gelatin mixture, yogurt, strawberries, salt and vanilla. Beat at medium speed until smooth. Pour into ice cream freezer container. Freeze using freezer directions. Serve immediately. Yield: 12 servings.
Note: Yogurt may be frozen in 3-quart metal mixer bowl instead of ice cream freezer as follows: Freeze yogurt mixture for 2 hours. Beat at medium speed for 1 minute. Repeat procedure and serve.

Approx per serving: Cal 68, Prot 3.2 gr,
T Fat 2.7 gr, Chol 46.8 mg, Carbo 7.8 gr,
Sod 59.6 mg, Potas 91.6 mg.

Nancy Szydelko
Sacramento, CA

Strawberries Royale

⅓ c. freshly squeezed orange juice
1⅓ c. sliced strawberries
½ c. raspberries

Combine orange juice and strawberries in bowl. Marinate for 1 to 4 hours. Purée raspberries with chopping blade in food processor. Press purée through fine sieve to remove seed. Drain strawberries. Spoon into dessert glasses. Top with raspberry sauce. Yield: 4 servings.

Approx per serving: Cal 36, Prot 0.7 gr,
T Fat 0.4 gr, Chol 0.0 mg, Carbo 8.4 gr,
Sod 0.9 mg, Potas 148.0 mg.

Julia Myers
Ft. Lauderdale, FL

Show-Me Strawberry Pie Dessert

1½ c. flour
½ c. oil
½ tsp. salt
2 tbsp. sugar
2 tbsp. milk
1 c. sugar
2 tbsp. white corn syrup
3 tbsp. cornstarch
1 c. water
2 tbsp. strawberry gelatin
4 c. small whole strawberries

Mix first 5 ingredients in bowl. Pat into 9x9-inch baking dish. Prick bottom with fork. Bake at 350 degrees until light brown. Cool. Mix 1 cup sugar, syrup, cornstarch and water in 3-quart saucepan. Cook over medium heat until thick and clear, stirring constantly. Add gelatin; stir until dissolved. Cool. Stir in strawberries. Spoon over crust. Chill until firm. Yield: 9 servings.

Approx per serving: Cal 344, Prot 3.2 gr,
T Fat 12.7 gr, Chol 0.5 mg, Carbo 55.8 gr,
Sod 136.1 mg, Potas 145.1 mg.

Debbie Hampton
Charleston Sr. H.S., Charleston, MO

Angel Food Cake

1½ c. egg whites
1½ tsp. cream of tartar
½ tsp. salt
1½ c. sugar
1 c. sifted cake flour
½ tsp. each vanilla, almond extract

Beat egg whites in bowl until foamy. Add cream of tartar and salt. Beat until medium-stiff peaks form. Fold in 1 cup sugar gently, using 25 strokes. Sift ½ cup sugar and flour together 3 times. Fold into egg whites ¼ at a time, using 15 strokes after each addition. Fold in flavoring, using 10 strokes. Spoon into ungreased 10-inch tube pan. Cut through batter with knife. Bake at 350 degrees for 45 minutes or until cake tests done. Invert on funnel to cool. Loosen edge with knife. Remove from pan. Yield: 16 servings.

Approx per serving: Cal 118, Prot 5.4 gr,
T Fat 0.0 gr, Chol 0.0 mg, Carbo 23.8 gr,
Sod 152.6 mg, Potas 79.6 mg.

Timmy Hickman
Blacktower, NM

Apple-Nut Cake

1 c. sugar
¼ c. shortening
1 egg
2 c. chopped unpeeled apple
1 c. sifted flour
1 tsp. each soda, cinnamon
½ tsp. nutmeg
½ tsp. salt
½ c. chopped pecans
½ c. raisins

Cream sugar and shortening in mixer bowl until light and fluffy. Blend in egg. Stir in apples. Add sifted dry ingredients; mix well. Stir in pecans and raisins. Pour into greased and floured 8x8-inch cake pan. Bake at 325 degrees for 45 minutes. Serve with whipped cream and maraschino cherry if desired. Yield: 9 servings.

Approx per serving: Cal 280, Prot 2.9 gr,
T Fat 11.8 gr, Chol 28.1 mg, Carbo 43.1 gr,
Sod 219.4 mg, Potas 151.6 mg.

Carol Brown
Fort Osage Jr. H.S., Independence, MO

Boiled Raisin Cake

1 15-oz. package raisins
2 c. water
¾ c. shortening
1½ c. sugar
½ tsp. salt
1 tsp. each cinnamon, nutmeg
½ tsp. each allspice, cloves
1 tbsp. soda
¼ c. hot coffee
2½ c. flour

Bring raisins and water to a boil in saucepan. Boil for 1 minute. Combine with shortening, sugar, salt and spices in bowl; mix well. Cool. Dissolve soda in coffee. Cool. Add flour to raisin mixture alternately with coffee; mixing well after each addition. Pour into greased and floured tube pan. Bake at 350 degrees until cake tests done. Cool in pan for 10 minutes. Remove to wire rack to cool completely. Yield: 20 servings.

Approx per serving: Cal 252, Prot 2.2 gr,
T Fat 8.6 gr, Chol 0.0 mg, Carbo 43.7 gr,
Sod 182.8 mg, Potas 183.4 mg.

Robin Bartoletti
Riverside Jr.-Sr. H.S., Dickson City, PA

Mystery Snack Cake

2 c. drained cooked pinto beans
¼ c. bean liquid
½ c. margarine, softened
⅔ c. sugar
1 egg
2 tsp. vanilla extract
1 c. whole wheat flour
½ c. instant nonfat dry milk powder
1½ tsp. baking powder
1 tsp. each cinnamon, allspice
1 c. raisins
½ c. chopped walnuts

Mash beans with bean liquid in bowl until smooth; set aside. Cream margarine and sugar in bowl until light and fluffy. Add egg, vanilla and beans; mix well. Add mixture of dry ingredients; mix well. Fold in raisins and walnuts. Pour into 9-inch square baking pan sprayed with nonstick cooking spray. Bake at 350 degrees for 1 hour or until cake tests done. Cool on wire rack. Garnish with sprinkle of confectioners' sugar.
Yield: 20 servings.

Approx per serving: Cal 149, Prot 3.1 gr,
T Fat 7.0 gr, Chol 13.0 mg, Carbo 20.1 gr,
Sod 95.4 mg, Potas 159.1 mg.

Prissy Berlin
Memphis, TN

Spicy Zucchini Cake

2 c. finely chopped zucchini
⅓ c. boiling water
2 c. flour
1¼ c. sugar
1¼ tsp. baking powder
1 tsp. salt
1 tsp. each cinnamon, nutmeg and cloves
½ c. oil
3 eggs
1 tsp. vanilla extract

Mix zucchini and boiling water in large mixer bowl. Add remaining ingredients. Beat at low speed for 1 minute, scraping bowl constantly. Beat at medium speed for 2 minutes, scraping bowl occasionally. Spoon into greased and floured 9x13-inch cake pan. Bake at 350 degrees for 45 to 50 minutes or until toothpick inserted in center comes out clean. Cool in pan on wire rack. Yield: 12 servings.

Approx per serving: Cal 261, Prot 4.1 gr,
T Fat 10.7 gr, Chol 63.2 mg, Carbo 37.6 gr,
Sod 228.0 mg, Potas 80.8 mg.

Fran Heckman
Waupaca H.S., Waupaca, WI

Frosted Carrot Bars

4 eggs
2 c. sugar
2 c. flour
1½ c. oil
1 tsp. salt
2 tsp. soda
2 tsp. cinnamon
3 c. grated carrots
1½ c. flaked coconut
1½ c. chopped walnuts
1 tbsp. milk
3 oz. cream cheese, softened
2½ c. confectioners' sugar
1 tsp. vanilla extract
⅛ tsp. salt
1 c. chopped walnuts
3 tbsp. milk

Beat eggs in mixer bowl until foamy. Add sugar gradually, beating until thick and lemon-colored. Add flour alternately with oil, mixing well after each addition. Mix in salt, soda and cinnamon. Fold in carrots, coconut and walnuts. Spread evenly in 2 greased 9x13-inch baking pans. Bake at 350 degrees for 30 minutes. Cool in pans on wire rack. Beat 1 tablespoon milk into cream cheese in small bowl. Add confectioners' sugar, vanilla, salt, walnuts and remaining 3 tablespoons milk; mix well. Spread on baked layers. Cut into 1x3-inch bars. Store in refrigerator or freezer. Yield: 78 bars.

Approx per bar: Cal 126, Prot 1.4 gr,
T Fat 7.9 gr, Chol 14.3 mg, Carbo 13.1 gr,
Sod 63.7 mg, Potas 45.5 mg.

Patricia Mikulecky
Bartlesville Mid.-H.S., Bartlesville, OK

Healthy Cookies

2 sticks unsalted butter, softened
1½ c. packed dark brown sugar
2 tsp. vanilla extract
2 eggs
2½ c. (or more) whole wheat flour
1½ tsp. soda
2 tsp. cinnamon
1 tsp. salt
1 c. oats
⅓ c. wheat germ
2 c. carob chips
⅓ c. coconut

Cream butter, brown sugar and vanilla in mixer bowl until light and fluffy. Add eggs 1 at a time, mixing well after each addition. Add flour, soda, cinnamon and salt; mix well. Stir in remaining ingredients. Drop by rounded teaspoonfuls 2 inches apart onto greased cookie sheets. Yield: 48 cookies.

Approx per cookie: Cal 131, Prot 1.9 gr,
T Fat 7.1 gr, Chol 23.3 mg, Carbo 16.8 gr,
Sod 76.7 mg, Potas 87.4 mg.

Judy Monroe
Hayes Mid. Sch., Grand Ledge, MI

Microwave Toffee Bars

1 c. quick-cooking oats
¼ c. packed brown sugar
¼ c. margarine, softened
2 tbsp. light corn syrup
¼ c. semisweet chocolate chips
3 tbsp. peanut butter

Mix oats, brown sugar, margarine and corn syrup in bowl. Spread evenly in 8x8-inch glass baking dish. Microwave on High for 4 minutes, turning occasionally. Place chocolate chips and peanut butter in glass measure. Microwave for 1 minute or until melted; mix well. Spread over oats layer. Chill in refrigerator. Cut into bars. Yield: 16 bars.

Approx per bar: Cal 95, Prot 1.6 gr,
T Fat 5.7 gr, Chol 0.0 mg, Carbo 10.4 gr,
Sod 54.4 mg, Potas 58.3 mg.

Sandra Whaley
North Whitfield Mid. Sch., Dalton, GA

Fortified Oatmeal Cookies

¾ c. margarine, softened
1½ c. packed brown sugar
1 egg
¼ c. water
1 tsp. vanilla extract
1 c. whole wheat flour
½ tsp. soda
1 tsp. salt
2½ c. oats
½ c. wheat germ
1 c. chopped pecans
½ c. sunflower seed

Beat margarine, brown sugar, egg, water and vanilla in mixer bowl until smooth. Add mixture of whole wheat flour, soda and salt; mix well. Mix in oats, wheat germ, pecans and sunflower seed. Drop by spoonfuls onto greased cookie sheets. Bake at 350 degrees for 12 to 15 minutes or until brown. Cool on wire rack. Yield: 60 cookies.

Approx per cookie: Cal 85, Prot 1.6 gr,
T Fat 4.7 gr, Chol 4.2 mg, Carbo 9.9 gr,
Sod 73.6 mg, Potas 70.3 mg.

Chris Ellingson
McClintock H.S., Tempe, AZ

Grandmother's Oatmeal-Coconut Cookies

1 c. shortening
1 c. each sugar, packed brown sugar
2 eggs
2 c. sifted flour
1 tsp. each baking powder, soda
½ tsp. salt
1 tsp. vanilla extract
2 c. quick-cooking oats
2 c. coconut

Cream shortening, sugar and brown sugar in mixer bowl until light and fluffy. Blend in eggs. Sift in flour, baking powder, soda and salt; mix well. Stir in vanilla, oats and coconut. Drop by teaspoonfuls onto greased cookie sheet. Bake at 350 degrees for 12 minutes. Remove to wire rack to cool. Yield: 36 cookies.

Approx per cookie: Cal 165, Prot 1.8 gr,
T Fat 8.4 gr, Chol 14.0 mg, Carbo 21.1 gr,
Sod 76.8 mg, Potas 61.5 mg.

Shari Rogers
Lincoln Mid. Sch., Abilene, TX

Raisin-Oatmeal Cookies

1 c. flour, sifted
½ tsp. soda
1 tsp. salt
¼ tsp. cinnamon
1½ c. quick-cooking oats
2 egg whites, slightly beaten
⅓ c. oil
1 c. packed brown sugar
½ c. skim milk
1 tsp. vanilla extract
1 c. raisins

Sift flour, soda, salt and cinnamon into bowl. Stir in oats. Combine egg whites, oil, brown sugar, milk, vanilla and raisins in bowl. Add to flour mixture; mix well. Drop by teaspoonfuls onto greased cookie sheet. Bake at 375 degrees for 12 minutes for chewy cookies or for 15 minutes for crisp cookies. Remove to wire rack to cool. Yield: 36 cookies.

Approx per cookie: Cal 80, Prot 1.3 gr,
T Fat 2.3 gr, Chol 0.1 mg, Carbo 14.1 gr,
Sod 48.5 mg, Potas 74.3 mg.

Janet Egbert
Carson City Crystal H.S., Saranac, MI

Honey Bunches

3 c. quick-cooking oats
2 c. flaked coconut
1 c. flour
1 c. butter
⅓ c. honey
1½ c. packed brown sugar

Mix oats, coconut and flour in large bowl. Bring butter, honey and brown sugar to a boil in saucepan, stirring constantly. Pour over dry ingredients; mix well. Spoon by tablespoonfuls into greased muffin cups. Bake at 350 degrees for 12 to 15 minutes or until well browned. Cool in muffin cups. Yield: 30 cookies.

Approx per cookie: Cal 177, Prot 1.8 gr,
T Fat 8.5 gr, Chol 18.9 mg, Carbo 24.4 gr,
Sod 90.2 mg, Potas 91.5 mg.

Carla Seippel
Fort Osage Jr. H.S., Independence, MO

Oatmeal-Granola Cookies

1 c. butter, softened
1 c. sugar
1 c. packed brown sugar
2 eggs, beaten
1 tsp. vanilla extract
1½ c. flour
1 tsp. each soda, salt
3 c. Granola

Cream butter, sugar and brown sugar in mixer bowl until light and fluffy. Blend in eggs and vanilla. Add flour, soda and salt; mix well. Stir in Granola. Shape into long rolls. Chill, wrapped in plastic wrap, for 1½ hours. Slice as desired. Place on cookie sheet. Bake at 375 degrees for 8 to 10 minutes or until brown. Cool on wire rack. Yield: 60 cookies.

Approx per cookie: Cal 100, Prot 1.3 gr,
T Fat 5.0 gr, Chol 17.9 mg, Carbo 13.2 gr,
Sod 91.1 mg, Potas 38.6 mg.

Granola

2   16-oz. packages oats
2 c. wheat germ
2 c. coconut
2 c. cashews
1½ c. oil
⅔ c. water
2 c. honey
3 tbsp. vanilla extract

Mix oats, wheat germ, coconut and cashews in bowl. Combine oil, water, honey and vanilla in saucepan. Bring to a boil. Stir into dry ingredients. Spread in shallow roasting pan. Bake at 350 degrees for 1 hour, stirring every 15 minutes. Yield: 18 cups.

Approx per cup: Cal 637, Prot 13.3 gr,
T Fat 33.0 gr, Chol 0.0 mg, Carbo 78.0 gr,
Sod 25.0 mg, Potas 397.3 mg.

Louann Heinrichs
Greenville H.S., Greenville, OH

Oatmeal-Wheat Crispies

1 c. shortening
1 c. sugar
1 c. packed brown sugar
2 eggs, well beaten
1 tsp. vanilla extract
1½ c. whole wheat flour
1 tsp. each soda, salt
3 c. oats
2 c. chopped walnuts

Cream shortening, sugar and brown sugar in mixer bowl until light and fluffy. Blend in eggs and vanilla. Add mixture of whole wheat flour, soda and salt; mix well. Stir in oats and walnuts. Shape into logs. Chill, wrapped in plastic wrap. Slice ¼ inch thick. Place on ungreased cookie sheet. Bake at 350 degrees for 10 minutes. Cool on wire rack. Yield: 60 cookies.
Note: May freeze logs before baking. May add ⅓ to ½ cup peanut butter or ¾ cup raisins, chocolate chips or chopped dried fruit.

Approx per cookie: Cal 114, Prot 1.8 gr,
T Fat 6.8 gr, Chol 8.4 mg, Carbo 12.4 gr,
Sod 52.6 mg, Potas 58.1 mg.

Katherine Anderson
East Central H.S., Lucedale, MS

Quaker Cookies

½ c. shortening
1 c. packed dark brown sugar
1 egg, beaten
2 tbsp. molasses
1 tsp. vanilla extract
1 c. flour
¾ tsp. soda
½ tsp. salt
¼ tsp. cinnamon
1½ c. oats
¼ c. golden raisins
¼ c. sugar

Cream shortening and brown sugar in mixer bowl until light and fluffy. Blend in egg, molasses and vanilla. Mix flour, soda, salt and cinnamon in small bowl. Add to creamed mixture; mix well. Stir in oats and raisins. Shape into 1-inch balls. Roll in sugar. Place on lightly greased cookie

sheet; flatten with thumb. Bake at 350 degrees for 10 to 12 minutes or until brown. Remove to wire rack to cool. Yield: 24 cookies.

Approx per cookie: Cal 134, Prot 1.6 gr,
T Fat 5.3 gr, Chol 10.5 mg, Carbo 20.6 gr,
Sod 76.3 mg, Potas 84.0 mg.

Kim Holloway
Sante Fe Trail Jr. H.S., Olathe, KS

Sugarless Orange Cookies

1 egg
½ c. margarine, softened
½ c. orange juice
1 tsp. grated orange rind
2 c. flour
2 tsp. baking powder
½ tsp. each cinnamon, salt
½ c. chopped pecans
½ c. raisins

Blend egg and margarine in mixer bowl. Add orange juice and orange rind; mix well. Add mixture of dry ingredients; beat until smooth. Stir in pecans and raisins. Drop by rounded teaspoonfuls 2 inches apart onto cookie sheet. Bake at 375 degrees for 15 to 20 minutes or until brown. Remove to wire rack to cool.
Yield: 24 cookies.

Approx per cookie: Cal 104, Prot 1.7 gr,
T Fat 5.9 gr, Chol 10.5 mg, Carbo 11.3 gr,
Sod 122.1 mg, Potas 62.4 mg.

Tamara Friesen
Jardine Jr. H.S., Wichita, KS

Cornflake Chews

1 c. light corn syrup
1 c. sugar
1 c. peanut butter
2 tbsp. margarine
1 tsp. vanilla extract
6 c. cornflakes

Heat corn syrup, sugar and peanut butter in saucepan over low heat, stirring until blended. Stir in margarine and vanilla. Add cornflakes; mix well. Drop by teaspoonfuls onto waxed paper.

Let stand until cool. Store in airtight container. Yield: 36 cookies.

Approx per cookie: Cal 151, Prot 2.1 gr, T Fat 4.3 gr, Chol 0.0 mg, Carbo 27.5 gr, Sod 93.3 mg, Potas 56.8 mg.

Loretta Briggs
Breckenridge Comm. Sch., Wheeler, MI

Healthy Honeys

1 c. chunky peanut butter
1 c. honey
1 c. nonfat dry milk powder
1½ c. crushed wheat Chex cereal
60 whole wheat Chex
2 c. crushed rice Chex cereal

Combine first 4 ingredients in bowl; mix well. Shape into 1-inch balls. Press 1 whole wheat Chex into each ball. Roll in crushed rice cereal. Chill, covered, for 2 hours. Yield: 60 snacks.

Approx per snack: Cal 60, Prot 1.8 gr, T Fat 2.2 gr, Chol 0.2 mg, Carbo 9.2 gr, Sod 60.8 mg, Potas 55.1 mg.

Lavern Frentzel
Perry Co. Dist. 32 Sch., Uniontown, MO

Microwave Naturally Good Treats

¾ c. natural unsalted peanut butter
¼ c. honey
3 tbsp. reduced-calorie margarine
½ c. nonfat dry milk powder
½ c. unsalted sunflower seed
⅓ c. sesame seed
⅓ c. crushed cornflakes
¼ c. chopped dried apricots
⅔ c. flaked coconut

Microwave peanut butter, honey and margarine in 1½-quart glass casserole on High until melted. Stir in next 5 ingredients. Shape into balls. Roll in coconut. Chill for 3 hours. Yield: 30 snacks.
Note: Nutritional information does not include reduced-calorie margarine.

Approx per snack: Cal 86, Prot 3.0 gr, T Fat 5.8 gr, Chol 0.2 mg, Carbo 6.4 gr, Sod 59.9 mg, Potas 102.4 mg.

Polly Anne Bryant
Lonoke Jr. H.S., Lonoke, AR

Nutritious Fig-Pecan Roll

1 tsp. grated orange rind
2 tbsp. fresh orange juice
1 c. chopped dried figs
½ c. miniature marshmallows
¼ c. butter
¼ c. chopped pecans
45 vanilla wafers, crushed
2 tsp. melted butter

Mix orange rind and orange juice in bowl. Add figs; set aside. Melt marshmallows with ¼ cup butter in saucepan over medium-high heat, stirring occasionally. Add fig mixture and pecans. Reserve ¼ cup vanilla wafer crumbs. Stir remaining crumbs into fig mixture; mix well. Brush 2 teaspoons melted butter onto 16-inch piece of waxed paper. Sprinkle with reserved crumbs. Spoon fig mixture onto prepared waxed paper. Shape into 9-inch roll, coating well. Chill, wrapped in waxed paper, for 12 hours. Cut into ¼-inch slices. Yield: 36 slices.

Approx per slice: Cal 48, Prot 0.4 gr, T Fat 3.3 gr, Chol 7.9 mg, Carbo 4.5 gr, Sod 36.2 mg, Potas 11.2 mg.

Sue P. Culpepper
Gulfport H.S., Gulfport, MS

Northern Plains Survival Logs

1 c. nonfat dry milk powder
1 c. honey
1 c. peanut butter
1 c. wheat germ
1 c. raisins
1 c. oatmeal, lightly toasted
½ c. chopped pecans
½ c. coconut
½ c. sesame seed
1 c. (or more) graham cracker crumbs

Combine all ingredients except graham cracker crumbs in large bowl; mix well. Shape into 2-inch logs. Roll in graham cracker crumbs. Yield: 24 logs.

Approx per log: Cal 239, Prot 7.2 gr, T Fat 12.0 gr, Chol 5.8 mg, Carbo 29.5 gr, Sod 120.9 mg, Potas 306.2 mg.

Mrs. Charles B. King
Decatur, TN

Peanut Butter Snacks

½ c. peanut butter
½ c. honey
1 c. toasted wheat germ
2 tbsp. nonfat dry milk powder
½ c. each raisins, coconut

Combine peanut butter, honey, wheat germ and dry milk powder in bowl; mix well. Stir in raisins. Shape into balls. Roll in coconut, coating well. Store, covered, in refrigerator. Yield: 15 snacks.

Approx per snack: Cal 137, Prot 4.5 gr,
T Fat 5.9 gr, Chol 0.1 mg, Carbo 19.1 gr,
Sod 62.5 mg, Potas 175.4 mg.

Janie Armstrong
Chippewa Valley H.S., Royal Oak, MI

Super Bites

1 c. honey
1 c. peanut butter
1 c. carob powder
1 c. each sesame seed, sunflower seed
½ c. each coconut, chopped dates
½ c. chopped pecans

Blend honey and peanut butter in bowl. Add remaining ingredients; mix well. Press into 8x8-inch pan. Chill until firm. Yield: 16 squares.

Approx per square: Cal 331, Prot 8.9 gr,
T Fat 21.0 gr, Chol 0.0 mg, Carbo 36.4 gr,
Sod 111.8 mg, Potas 329.0 mg.

Judy Monroe
Hayes Mid. Sch., Grand Ledge, MI

Sweet Cereal Puffs

3 egg whites
⅔ c. sugar
4 c. Total cereal

Beat egg whites in bowl until foamy. Add sugar gradually, beating until stiff. Fold in cereal. Drop by teaspoonfuls 2 inches apart on greased cookie sheet. Bake at 325 degrees for 14 minutes or until brown. Cool on wire rack. Yield: 42 puffs.

Approx per puff: Cal 25, Prot 0.6 gr,
T Fat 0.1 gr, Chol 0.0 mg, Carbo 5.7 gr,
Sod 42.4 mg, Potas 15.1 mg.

Donna Dyess
Crystal Springs H.S., Crystal Springs, MS

Kiwi Chiffon Pie

¾ c. flour
¼ tsp. salt
8 tsp. reduced-calorie margarine
¼ c. low-fat yogurt
1   2-serving env. reduced-calorie
*   lemon pudding mix*
¾ c. water
2 egg whites, at room temperature
1 tbsp. sugar
1 med. kiwifruit, peeled, thinly sliced

Mix flour and salt in bowl. Cut in margarine until crumbly. Stir in yogurt; shape into ball. Chill, wrapped in plastic wrap, for 1 hour to 3 days. Roll to 10-inch circle between waxed paper. Chill for several minutes. Fit into 8-inch pie plate. Trim and flute edge; prick with fork. Bake at 400 degrees for 15 minutes or until light brown. Cool on wire rack. Combine pudding mix and water in saucepan. Bring to a boil over medium heat, stirring constantly. Chill, covered, until slightly thickened, stirring occasionally. Beat egg whites until soft peaks form. Add sugar gradually, beating until stiff peaks form. Fold into pudding. Spoon into pie shell. Arrange kiwifruit slices over top. Chill, covered, until serving time.
Yield: 8 servings.
Note: Nutritional information does not include kiwifruit slices .

Approx per serving: Cal 90, Prot 3.4 gr,
T Fat 3.3 gr, Chol 0.0 mg, Carbo 13.7 gr,
Sod 159.1 mg, Potas 33.6 mg.

Karen Blanton
Hendersonville, TN

Lemon Cheesecake Pie

3 tbsp. melted margarine
1¼ c. graham cracker crumbs
1   3-oz. package lemon gelatin
1 c. boiling water
2 c. low-fat cottage cheese
1 to 2 tsp. grated lemon rind

Combine margarine and graham cracker crumbs in bowl; mix well. Press over bottom and side of 9-inch pie plate. Bake at 350 degrees for 4 to 5 minutes or until light brown. Cool. Chill in refrigerator until firm. Dissolve gelatin in boiling water in bowl; cool to lukewarm. Combine cottage cheese and lemon rind in blender container. Process until smooth. Add to gelatin gradually, mixing well. Pour into pie shell. Chill until set. Yield: 8 servings.

Note: Nutritional information does not include low-fat cottage cheese.

Approx per serving: Cal 141, Prot 2.4 gr,
T Fat 5.9 gr, Chol 0.0 mg, Carbo 21.5 gr,
Sod 196.3 mg, Potas 87.3 mg.

Sharie Mueller
Jefferson Co. North H.S., Oskaloosa, KS

Lemon Chess Pie

4 eggs
¼ c. milk
1¼ c. sugar
Grated rind of 1 lemon
1 tbsp. flour
1 tbsp. cornmeal
¼ c. melted butter
1 unbaked 9-in. pie shell
½ tsp. nutmeg

Beat eggs, milk, sugar and lemon rind in mixer bowl until smooth. Add flour and cornmeal; mix well. Stir in butter. Pour into pie shell. Sprinkle with nutmeg. Bake at 350 degrees for 45 to 50 minutes or until set. Yield: 6 servings.

Approx per serving: Cal 450, Prot 6.8 gr,
T Fat 21.9 gr, Chol 193.6 mg, Carbo 57.6 gr,
Sod 323.5 mg, Potas 80.5 mg.

Kathleen Burchett
Area Supr. of H.E., State Dept. of Ed., Abingdon, VA

Lemon Pie Meringue

2 egg whites
½ tsp. vinegar
½ tsp. vanilla extract
⅛ tsp. salt
3 tbsp. sugar
1 tbsp. finely shredded lemon rind
2 eggs
¼ c. lemon juice
¼ c. sugar
1 tbsp. butter
1 env. whipped topping mix
½ c. low-fat milk
½ tsp. vanilla extract

Spray 9-inch pie plate with nonstick cooking spray. Combine egg whites and next 3 ingredients in small mixer bowl. Beat until soft peaks form. Add 3 tablespoons sugar gradually, beating until stiff peaks form. Spread evenly over bottom and side of prepared pie plate to form crust. Bake at 325 degrees for 25 minutes or until light brown. Cool. Combine lemon rind, eggs, lemon juice, ¼ cup sugar and butter in saucepan. Cook over low heat until thickened, stirring constantly. Spread ¼ of the mixture over meringue. Prepare whipped topping mix according to package directions using low-fat milk and ½ teaspoon vanilla. Spread half the mixture in pie plate. Fold remaining lemon mixture into whipped topping. Spoon over top. Chill, covered, in refrigerator. Garnish with lemon twists. Yield: 8 servings.

Approx per serving: Cal 89, Prot 3.1 gr,
T Fat 3.1 gr, Chol 68.2 mg, Carbo 12.6 gr,
Sod 87.4 mg, Potas 64.4 mg.

Lonnie Peters
Mt. Pleasant, TX

Fruit Pizza

1 pkg. sugar cookie mix
1 egg
¼ c. water
8 oz. cream cheese, softened
⅓ c. sugar
1 8-oz. can crushed pineapple
1 banana, sliced
1 8-oz. can mandarin oranges, drained
18 maraschino cherries
1 2-oz. package walnut chips
½ c. shredded coconut

Combine cookie mix, egg and water in bowl; mix well. Spread in pizza pan. Bake at 375 degrees for 10 to 12 minutes or until brown. Cool. Beat cream cheese and sugar in mixer bowl until light. Spread over cooled crust. Drain pineapple, reserving juice. Dip banana slices into reserved juice to prevent discoloring; drain. Arrange banana slices, pineapple and remaining ingredients on cream cheese layer. Cut into wedges. Yield: 6 servings.

Note: Nutritional information does not include cookie mix.

Approx per serving: Cal 380, Prot 6.3 gr,
T Fat 23.6 gr, Chol 84.1 mg, Carbo 39.2 gr,
Sod 120.9 mg, Potas 254.2 mg.

Elizabeth Wintersteen
Central-Columbia H.S., Danville, PA

Cream Cheese Tarts

24 oz. cream cheese, softened
5 eggs
1 c. sugar
1½ tsp. vanilla extract
2 c. sour cream
½ c. sugar
1 can cherry pie filling

Beat cream cheese in bowl. Add eggs, 1 at a time, mixing well after each addition. Mix in 1 cup sugar and vanilla. Fill paper-lined miniature muffin cups ¾ full. Bake at 325 degrees for 20 minutes. Cool for 5 minutes. Blend sour cream with remaining ½ cup sugar. Spoon into center of each tart. Bake for 5 minutes longer. Cool. Spoon pie filling over top. Chill in refrigerator. Yield: 24 servings.

Approx per serving: Cal 147, Prot 2.3 gr,
T Fat 6.6 gr, Chol 65.0 mg, Carbo 20.2 gr,
Sod 31.0 mg, Potas 44.0 mg.

Mary Lou Simms
Mattoon, IL

Summer Pear Tart

¾ c. flour
6 tbsp. butter
3 tbsp. sugar
¼ tsp. allspice
1 env. unflavored gelatin
½ c. orange juice
3 tbsp. brown sugar
¼ tsp. salt
⅛ tsp. allspice
¼ c. orange marmalade
3 ripe med. pears, puréed
3 ripe med. pears, thinly sliced
¼ c. orange marmalade, melted

Combine first 4 ingredients in bowl. Knead until blended. Pat over bottom and side of 9-inch springform tart pan. Prick with fork. Bake at 400 degrees for 10 minutes or until golden. Cool on wire rack. Soften gelatin in orange juice in saucepan for 1 minute. Cook over low heat until dissolved, stirring constantly. Remove from heat. Stir in next 4 ingredients and pear purée. Chill for 40 minutes, stirring occasionally. Pour into crust. Chill, covered, for 2 hours. Arrange sliced pears over top of tart. Brush with melted marmalade. Remove side of pan. Yield: 12 servings.

Approx per serving: Cal 194, Prot 2.1 gr,
T Fat 6.2 gr, Chol 17.7 mg, Carbo 35.2 gr,
Sod 119.8 mg, Potas 152.2 mg.

D'Lyn Loessin
Nueces County, TX

Dieter's Pie Crust

½ c. cottage cheese, drained
2 tbsp. shortening
⅓ c. sifted flour
⅛ tsp. salt

Squeeze cottage cheese with towel until dry. Force through fine sieve. Cut shortening into mixture of flour and salt in bowl. Mix in cottage cheese with fork to form ball. Roll on lightly floured pastry cloth. Fit into pie plate. Bake at 400 degrees for 20 minutes. Yield: 1 pie shell.

Approx per pie shell: Cal 516, Prot 20.6 gr,
T Fat 33.5 gr, Chol 23.3 mg, Carbo 32.4 gr,
Sod 558.4 mg, Potas 140.2 mg.

Sarah Briley
Lawrence, KS

Strawberry-Orange Leather

1 20-oz. package frozen whole
strawberries, thawed, drained
1 med. orange, peeled, sliced
2 tbsp. sugar
1½ to 3 tsp. lemon juice

Line bottoms of two 10x15-inch baking pans with plastic wrap. Place strawberries and orange in blender container. Process until smooth. Add sugar and lemon juice; mix well. Spread evenly in prepared pans. Place in 150-degree oven, leaving door open 2 inches. Bake for 3 hours or until dry. Invert onto cutting board; remove plastic wrap. Cut each sheet crosswise into 7 strips. Roll strips in plastic wrap. Store at room temperature. Yield: 14 servings.

Approx per serving: Cal 21, Prot 0.4 gr,
T Fat 0.2 gr, Chol 0.0 mg, Carbo 4.9 gr,
Sod 0.4 mg, Potas 73.3 mg.

Fran Heckman
Waupaca H.S., Waupaca, WI

COUNT YOUR CALORIES

Almonds, shelled, ¼ cup 213
Apples: 1 med. 70
   chopped, ½ cup 30
   juice, 1 cup 117
Applesauce: sweet, ½ cup 115
   unsweetened, ½ cup 50
Apricots: fresh, 3 55
   canned, ½ cup 110
   dried, 10 halves 100
   nectar, 1 cup 140
Asparagus: fresh, 6 spears 19
   canned, ½ cup 18
Avocado, 1 med. 265
Bacon, crisp-cooked, 2 sl. 90
Banana, 1 med. 100
Beans: baked, ½ cup 160
   dry, ½ cup 350
   green, ½ cup 20
   lima, ½ cup 95
   soy, ½ cup 95
Bean sprouts, ½ cup 18
Beef, cooked, 3 oz:
   broiled, sirloin steak 330
   roasted, heel of round 165
   roasted, rib 375
Beer, 12 oz. 150
Beets, cooked, ½ cup 40
Biscuit, from mix, 1 90
Bologna, all meat, 3 oz. 235
Bread: roll, 1 85
   white, 1 slice 65
   whole wheat, 1 slice 55
Bread crumbs, dry, 1 cup 390
Broccoli, cooked, ½ cup 20
Butter, 1 tbsp. 100
Buttermilk, 1 cup 90
Cabbage: cooked, ½ cup 15
   fresh, shredded, ½ cup 10
Cake: angel food, 1/12 140
   devil's food, 1/12 195
   yellow, 1/12 200
Candy: caramel, 1 oz. 115
   chocolate, sweet, 1 oz. 145
   hard candy, 1 oz. 110
   marshmallows, 1 oz. 90
Cantaloupe, ½ med. 60
Carrots: cooked, ½ cup 23
   fresh, 1 med. 20
Catsup, 1 tbsp. 18
Cauliflower: cooked, ½ cup 13
   fresh, ½ lb. 60
Celery, chopped, ½ cup 8
Cereals: bran flakes, ½ cup 53

cornflakes, ½ cup 50
   oatmeal, cooked, ½ cup 65
Cheese: American, 1 oz. 105
   Cheddar, 1 oz. 113
   cottage: creamed, ½ cup 130
     uncreamed, ½ cup 85
   cream, 1 oz. 107
   mozzarella, 1 oz. 80
   Parmesan, 1 oz. 110
   Velveeta, 1 oz. 84
Cherries: canned, sour, ½ cup 53
   fresh, sweet, ½ cup 40
Chicken, cooked, 4 oz:
   broiled 155
   canned, boned 230
   roasted, dark meat 210
   roasted, light meat 207
Chilies: green, fresh, ½ lb. 62
   red, fresh, ½ lb. 108
Chili powder, 1 tbsp. 51
Chocolate, baking, 1 oz. 143
Cocoa mix, 1-oz. package 115
Cocoa powder, ⅓ cup 120
Coconut, shredded, ¼ cup 166
Coffee 0
Corn: cream-style, ½ cup 100
   whole kernel, ½ cup 85
Corn bread, 1x4-in. piece 125
Corn chips, 1 oz. 130
Cornmeal, ½ cup 264
Cornstarch, 1 tbsp. 29
Crab meat: fresh, 3 oz. 80
   canned, 3 oz. 85
Crackers: graham, 1 square 28
   Ritz, each 17
   saltine, 2-in. square 13
Cracker crumbs, ½ cup 281
Cranberries: fresh, ½ lb. 100
   juice, cocktail, 1 cup 163
   sauce, ½ cup 190
Cream: half-and-half, 1 tbsp. 20
   heavy, 1 tbsp. 55
   light, 1 tbsp. 30
Creamer, nondairy powder, 1 tsp. 10
Cucumber, 1 med. 30
Dates, chopped, ½ cup 244
Eggs: 1 whole, large 80
   1 white 17
   1 yolk 59
Eggplant, cooked, ½ cup 19
Fish sticks, 5 200
Flour: all-purpose: 1 cup 420
   1 tbsp. 28

rye, 1 cup	286
whole wheat, 1 cup	400
Fruit cocktail, canned, ½ cup	98
Garlic, 1 clove	2
Gelatin, unflavored, 1 env.	25
Grapes: fresh, ½ cup	35-50
juice, 1 cup	170
Grapefruit: fresh, ½ med.	60
juice, unsweetened, 1 cup	100
Ground beef, cooked, 3 oz:	
lean	185
regular	245
Haddock, fried, 3 oz.	140
Ham, cooked, 3 oz:	
boiled	200
country-style	335
cured, lean	160
roasted, fresh,	320
Honey, 1 tbsp.	65
Ice cream, ½ cup	135
Ice milk, ½ cup	96
Jams and preserves, 1 tbsp.	54
Jellies, 1 tbsp.	55
Jell-O, ½ cup	80
Lamb, cooked, 3 oz:	
broiled, rib chop	175
roasted, leg,	185
Lemonade, sweetened, 1 cup	110
Lemon juice, 1 tbsp.	4
Lentils, cooked, ½ cup	168
Lettuce, 1 head	40
Liver: beef, fried, 2 oz.	130
chicken, simmered, 2 oz.	88
Lobster, 2 oz.	55
Macaroni, cooked, ½ cup	90
Mango, 1 fresh	134
Margarine, 1 tbsp.	100
Mayonnaise, 1 tbsp.	100
Milk: whole, 1 cup	160
skim, 1 cup	89
2-percent, 1 cup	145
condensed, 1 cup	982
evaporated, 1 cup	385
nonfat dry, 1 cup	251
Mushrooms: canned, ½ cup	20
fresh, 1 lb.	123
Mustard: prepared, brown, 1 tbsp.	13
prepared, yellow, 1 tbsp.	10
Nectarine, 1 fresh	30
Noodles: egg, cooked, ½ cup	100
fried, chow mein, 2 oz.	275
Oil, cooking, salad, 1 tbsp.	120
Okra, cooked, 8 pods	25
Olives: green, 3 lg.	15
ripe, 2 lg.	15

Onion: chopped, ½ cup	32
dehydrated flakes, 1 tbsp.	17
green, 6	20
Oranges: 1 whole	65
juice, 1 cup	115
Oysters, ½ cup	80
Peaches: fresh, 1 med.	35
canned, ½ cup	100
dried, ½ cup	210
Peanuts, roasted, 1 cup	420
Peanut butter, 1 tbsp.	100
Pears: fresh, 1 med.	100
canned, ½ cup	97
dried, ½ cup	214
Peas: black-eyed, ½ cup	70
green, canned, ½ cup	83
green, frozen, ½ cup	69
Pecans, chopped, ½ cup	400
Peppers: sweet green, 1 med.	14
sweet red, 1 med.	19
Perch, white, 4 oz.	50
Pickles: dill, 1 lg.	15
sweet, 1 avg.	30
Pie, ⅙ pie:	
apple	420
cherry	402
custard	330
pumpkin	321
Pie crust, mix, 1 crust	626
Pimento, canned, 1 avg.	10
Pineapple: fresh, ½ cup	36
canned, ½ cup	90
juice, 1 cup	135
Plums: fresh, 1 med.	30
canned, 3	101
Popcorn: plain, popped, 1 cup	23
with oil and salt, 1 cup	40
Pork, cooked, lean:	
broiled, chop, 3.5 oz.	260
roasted, Boston Butt, 4 oz.	280
roasted, loin, 4 oz.	290
Potato chips, 10 med.	114
Potatoes, white:	
baked, 1 sm. with skin	93
boiled, 1 sm.	70
French-fried, 10 pieces	175
hashed brown, ½ cup	177
mashed, ½ cup	90
Potatoes, sweet:	
baked, 1 avg.	155
candied, 1 avg.	295
canned, ½ cup	110
Prunes: 1 lg.	19
dried, cooked, ½ cup	137
juice, 1 cup	197

Puddings, instant, prepared, ½ cup:
    banana 175
    butterscotch 175
    chocolate 200
    lemon 180
Puddings, pie fillings, prepared, ½ cup:
    banana 165
    butterscotch 190
    chocolate 190
    lemon 125
Pumpkin, canned, ½ cup 38
Raisins, dried, ½ cup 231
Rice: cooked, white, ½ cup 90
    cooked, brown, ½ cup 100
    precooked, ½ cup 105
Salad dressings, commercial, 1 tbsp:
    blue cheese 75
    French 70
    Italian 83
    mayonnaise 100
    mayonnaise-type 65
    Russian 75
    Thousand Island 80
Salami, cooked, 2 oz. 180
Salmon: canned, 4 oz. 180
    steak, 4 oz. 220
Sardines, canned, 3 oz. 75
Sauces: barbecue, 1 tbsp. 17
    hot pepper, 1 tbsp. 3
    soy, 1 tbsp. 9
    Tartar, 1 tbsp. 74
    white, med., ½ cup 215
    Worcestershire, 1 tbsp. 15
Sauerkraut, ½ cup 21
Sausage, cooked, 2 oz. 260
Sherbet, ½ cup 130
Shrimp: cooked, 3 oz. 50
    canned, 4 oz. 130
Soft drinks, 1 cup 100
Soup, condensed, 1 can:
    chicken with rice 116
    cream of celery 215
    cream of chicken 235
    cream of mushroom 331
    tomato 220
    vegetable-beef 198

Sour cream, ½ cup 240
Spaghetti, cooked, ½ cup 80
Spinach: fresh, ½ lb. 60
    cooked, ½ cup 20
Squash: summer, ½ cup 15
    winter, ½ cup 65
Strawberries, fresh, ½ cup 23
Sugar: brown, ½ cup 410
    confectioners', ½ cup 240
    granulated, ½ cup 385
      1 tbsp. 48
Syrups: chocolate, 1 tbsp. 50
    corn, 1 tbsp. 58
    maple, 1 tbsp. 50
Taco shell, 1 shell 50
Tomatoes: fresh, 1 med. 40
    canned, ½ cup 25
    juice, 1 cup 45
    paste, 6-oz. can 150
    sauce, 8-oz. can 34
Toppings: caramel, 1 tbsp. 70
    chocolate fudge, 1 tbsp. 65
    Cool Whip, 1 tbsp. 14
    Dream Whip, 1 tbsp. 8
    strawberry, 1 tbsp. 60
Tortilla, corn, 1 65
Tuna: canned in oil, 4 oz. 230
    canned in water, 4 oz. 144
Turkey, roasted, 4 oz:
    dark 230
    light 200
Veal, cooked, 3 oz:
    broiled, cutlet 185
    roast 230
Vegetable juice cocktail, 1 cup 43
Vinegar, 1 tbsp. 2
Waffles, 1 130
Walnuts, chopped, ½ cup 410
Water chestnuts, ½ cup 25
Watermelon, fresh, ½ cup 26
Wheat germ, 1 tbsp. 29
Yeast: cake, 1 oz. 24
    dry, 1 oz. 80
Yogurt: plain, 1 cup 153
    plain, skim milk, 1 cup 123
    with fruit, 1 cup 260

NUTRIENTS FOR GOOD HEALTH

The food we eat should supply the nutrients we need for health and growth. Each nutrient has a certain job to do in the body, and it is important to understand the functions of the different nutrients because many of them work together.

The best way to get all the necessary nutrients is to eat a variety of foods. This variety supplies a balance of nutrients and minimizes the possibility of overloading the system with any one nutrient. Most foods contain more than one nutrient, but no one food provides all the different nutrients needed.

SOME IMPORTANT NUTRIENTS

PROTEIN
Why?   Builds and maintains all tissues. Forms an important part of enzymes, hormones and body fluids. Helps form antibodies to fight infection. Supplies energy.

Where?   Milk and cheese; poultry, fish and shellfish; cereals and breads; nuts and dried beans and peas; vegetables.

CARBOHYDRATES
Why?   Supply energy freeing protein for use in building and repairing body tissues. Help body use fat efficiently.

Where?   Starches: breads, cereals and rice; pastas; vegetables such as corn and potatoes.
Sugars: honey, jams, jellies, fruits, molasses, sugars and syrups.

FATS
Why?   Supply essential fatty acids. Help body use fat soluble vitamins A, D, E and K. Contain concentrated source of energy.

Where?   Butter, margarine, cream, whole milk and cheese; oils; meat, fish and poultry; nuts; chocolate.

VITAMIN A
Why?   Helps keep skin healthy. Helps keep lining of mouth, nose, throat and digestive tract healthy and resistant to infection. Promotes growth. Helps eyes adjust to dim light.

Where?   Butter and fortified margarine; liver; dark yellow fruits such as apricots and cantaloupes; dark green leafy and yellow vegetables such as spinach, chard, broccoli, red leaf lettuce, carrots, pumpkin and sweet potatoes.

THIAMIN (VITAMIN B$_1$)
Why?   Helps cells use energy from foods. Helps keep nerves in healthy condition. Promotes good appetite and digestion.

Where?   Lean pork and liver; whole grain or enriched breads and cereals; nuts and dried beans and peas.

RIBOFLAVIN (VITAMIN B$_2$)

Why? Helps cells use oxygen to release energy from food. Helps keep eyes healthy. Helps keep skin around mouth and nose smooth.

Where? Milk and cheese; eggs; meat, especially liver and kidney; enriched breads and cereals; dried beans and peas; dark green leafy vegetables.

NIACIN

Why? Helps cells use oxygen to produce energy. Helps maintain healthy skin, tongue, digestive tract and nervous system.

Where? Lean meat, fish, poultry and liver; whole grain or enriched breads and cereals; peanut butter, beans and peas.

VITAMIN B$_{12}$

Why? Helps form normal red blood cells.

Where? Milk and cheese; eggs; meat, fish and poultry.

VITAMIN C (ASCORBIC ACID)

Why? Helps build bones and teeth. Strengthens walls of blood vessels. Helps in healing wounds.

Where? Citrus fruits and strawberries; broccoli, cabbage, green and red peppers, potatoes and tomatoes.

VITAMIN D

Why? Helps use calcium and phosphorus to build strong bones and teeth.

Where? Vitamin D-fortified milk; fish liver oils; direct sunlight.

IRON

Why? Helps form red blood cells. Carries oxygen to body cells so energy can be released from foods.

Where? Red meat, fish and poultry; whole grain or enriched breads and cereals; dried beans and peas; green leafy vegetables.

CALCIUM

Why? Builds strong bones and teeth. Regulates heart. Helps muscle contraction, nerve transmission and normal blood clotting.

Where? Milk and milk products except butter; oysters, clams and sardines; dark green vegetables.

IODINE

Why? Helps thyroid gland work properly.

Where? Iodized salt; salt water fish and other seafood.

Substitution Chart

	Instead of:	Use:
BAKING	1 teaspoon baking powder 1 tablespoon cornstarch (for thickening) 1 cup sifted all-purpose flour 1 cup sifted cake flour	¼ teaspoon soda plus ½ teaspoon cream of tartar 2 tablespoons flour or 1 tablespoon tapioca 1 cup plus 2 tablespoons sifted cake flour 1 cup minus 2 tablespoons sifted all-purpose flour
	1 cup fine dry bread crumbs	¾ cup fine cracker crumbs
DAIRY	1 cup buttermilk 1 cup heavy cream 1 cup light cream 1 cup sour cream 1 cup sour milk	1 cup sour milk or 1 cup yogurt ¾ cup skim milk plus ⅓ cup butter ⅞ cup skim milk plus 3 tablespoons butter ⅞ cup sour milk plus 3 tablespoons butter 1 cup sweet milk plus 1 tablespoon vinegar or lemon juice or 1 cup buttermilk
SEASONINGS	1 teaspoon allspice 1 cup catsup 1 clove of garlic 1 teaspoon Italian spice 1 teaspoon lemon juice 1 tablespoon prepared mustard 1 medium onion	½ teaspoon cinnamon plus ⅛ teaspoon cloves 1 cup tomato sauce plus ½ cup sugar plus 2 tablespoons vinegar ⅛ teaspoon garlic powder or ⅛ teaspoon instant minced garlic or ¾ teaspoon garlic salt or 5 drops of liquid garlic ¼ teaspoon each oregano, basil, thyme, rose- mary plus dash of cayenne ½ teaspoon vinegar 1 teaspoon dry mustard 1 tablespoon dried minced onion or 1 teaspoon onion powder
SWEET	1 1-ounce square chocolate 1⅔ ounces semisweet chocolate 1 cup honey 1 cup granulated sugar	3 to 4 tablespoons cocoa plus 1 teaspoon shortening 1 ounce unsweetened chocolate plus 4 tea- spoons granulated sugar 1 to 1¼ cups sugar plus ¼ cup liquid or 1 cup corn syrup or molasses 1 cup packed brown sugar or 1 cup corn syrup, molasses or honey minus ¼ cup liquid

Equivalent Chart

	When the recipe calls for:	You need:
BAKING	½ cup butter	1 stick
	2 cups butter	1 pound
	4 cups all-purpose flour	1 pound
	4½ to 5 cups sifted cake flour	1 pound
	1 square chocolate	1 ounce
	1 cup semisweet chocolate pieces	1 6-ounce package
	4 cups marshmallows	1 pound
	2¼ cups packed brown sugar	1 pound
	4 cups confectioners' sugar	1 pound
	2 cups granulated sugar	1 pound
	3 cups tapioca	1 pound
BREAD & CEREAL	1 cup fine dry bread crumbs	4 to 5 slices
	1 cup soft bread crumbs	2 slices
	1 cup small bread cubes	2 slices
	1 cup fine cracker crumbs	28 saltines
	1 cup fine graham cracker crumbs	15 crackers
	1 cup vanilla wafer crumbs	22 wafers
	1 cup crushed cornflakes	3 cups uncrushed
	4 cups cooked macaroni	1 8-ounce package
	3½ cups cooked rice	1 cup uncooked
DAIRY	1 cup freshly grated cheese	¼ pound
	1 cup cottage cheese	1 8-ounce carton
	1 cup sour cream	1 8-ounce carton
	1 cup whipped cream	½ cup heavy cream
	⅔ cup evaporated milk	1 small can
	1⅔ cups evaporated milk	1 13-ounce can
FRUIT	4 cups sliced or chopped apples	4 medium
	1 cup mashed banana	3 medium
	2 cups pitted cherries	4 cups unpitted
	3 cups shredded coconut	½ pound
	4 cups cranberries	1 pound
	1 cup pitted dates	1 8-ounce package
	1 cup candied fruit	1 8-ounce package
	3 to 4 tablespoons lemon juice plus 1 teaspoon grated rind	1 lemon
	⅓ cup orange juice plus 2 teaspoons grated rind	1 orange
	4 cups sliced peaches	8 medium
	2 cups pitted prunes	1 12-ounce package
	3 cups raisins	1 15-ounce package

When the recipe calls for:	You need:
MEAT	
4 cups chopped cooked chicken	1 5-pound chicken
3 cups chopped cooked meat	1 pound, cooked
2 cups cooked ground meat	1 pound, cooked
1 cup chopped nuts	4 ounces, shelled
	1 pound, unshelled
VEGETABLES	
2 cups cooked green beans	½ pound fresh
	or 1 16-ounce can
2½ cups lima beans or red beans	1 cup dried, cooked
4 cups shredded cabbage	1 pound
1 cup grated carrots	1 large
1 4-ounce can mushrooms	½ pound, fresh
1 cup chopped onion	1 large
4 cups sliced or chopped raw potatoes	4 medium
2 cups canned tomatoes	1 16-ounce can

Common Equivalents

1 tablespoon = 3 teaspoons
2 tablespoons = 1 ounce
4 tablespoons = ¼ cup
5 tablespoons + 1 teaspoon
    = ⅓ cup
8 tablespoons = ½ cup
12 tablespoons = ¾ cup
16 tablespoons = 1 cup
1 cup = 8 ounces or ½ pint
4 cups = 1 quart
4 quarts = 1 gallon

6½ to 8-ounce can = 1 cup
10½ to 12-ounce can = 1¼ cups
14 to 16-ounce can (No. 300) = 1¾ cups
16 to 17-ounce can (No. 303) = 2 cups
1-pound 4-ounce can or 1-pint 2-ounce can
    (No. 2) = 2½ cups
1-pound 13-ounce can (No. 2½) = 3½ cups
3-pound 3-ounce can or 46-ounce can
    = 5¾ cups
6½-pound or 7-pound 5-ounce can (No. 10)
    = 12 to 13 cups

Metric Conversion Chart

Liquid		**Dry**	
1 teaspoon = 5 milliliters		1 quart = 1 liter	
1 tablespoon = 15 milliliters		1 ounce = 30 grams	
1 fluid ounce = 30 milliliters		1 pound = 450 grams	
1 cup = 250 milliliters		2.2 pounds = 1 kilogram	
1 pint = 500 milliliters			

NOTE: The metric measures are approximate benchmarks for purposes of home food preparation.

INDEX

FAVORITE RECIPES PRESS COOKBOOKS

For Your Collection

or

As A Gift

FOR ORDERING INFORMATION

Write to:

Favorite Recipes Press
P.O. Box 305142, Nashville, TN   37230

or

Call Toll-free
1-800-251-1542